I0901029

IMAGES
of America

FIREFIGHTING IN BUNCOMBE COUNTY

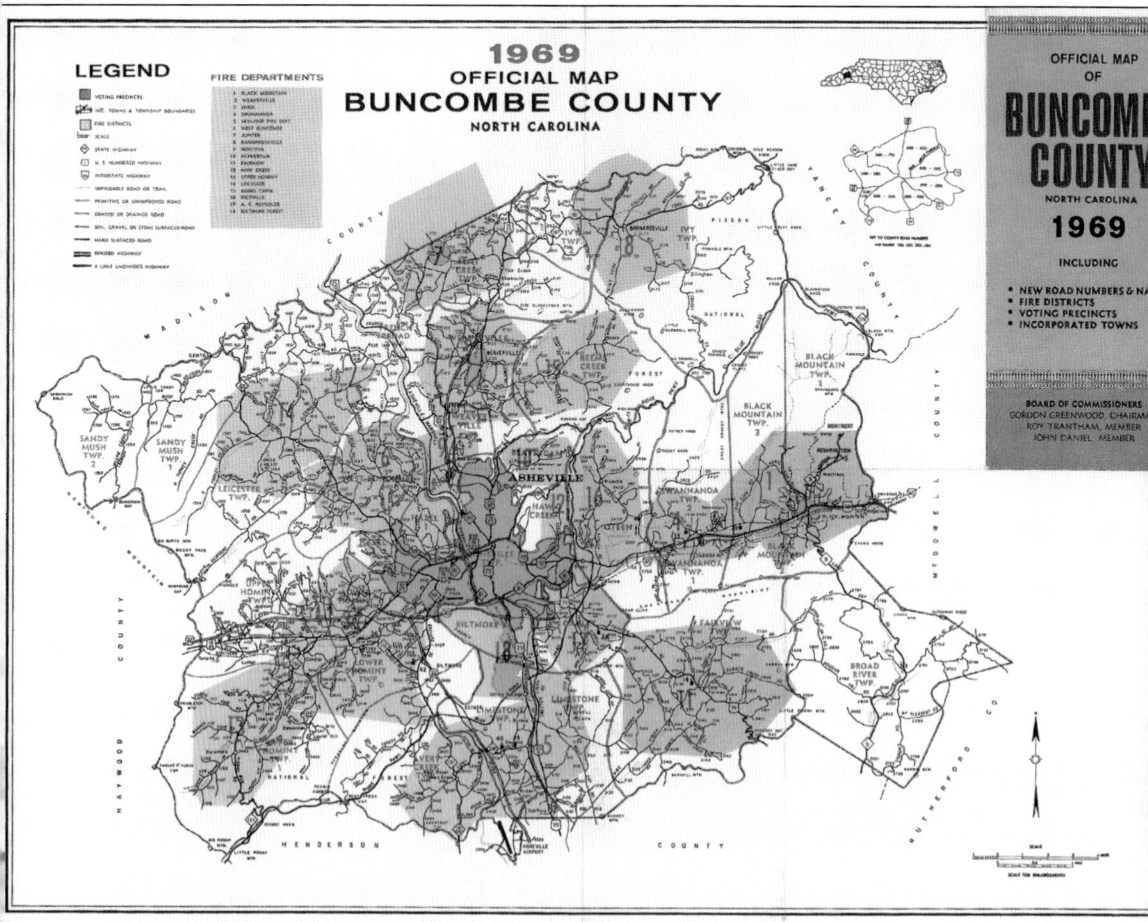

This map shows the 656-square-mile area of Buncombe County that will be covered in this book. Several fire districts are shown that were formed by 1969, but more fire departments and other agencies in the county will be covered. Although flat on this page, Buncombe County is anything but, with a widely varied topography presenting interesting challenges to emergency response and tactics. (Courtesy of North Carolina Collection, Pack Memorial Public Library, Asheville.)

On the Cover: This photograph shows a uniformed Biltmore Forest dual police officer and firefighter standing in front of their first fire truck. The building behind is the first town hall for the 1923-incorporated town. Their fire department and at least 14 others in the county no longer exist. They all will be discussed with what information is available. (Courtesy of E.M. Ball Collection, Special Collections, Ramsey Library, UNCA.)

IMAGES *of America*

Firefighting in Buncombe County

Brian Lawrence

ARCADIA PUBLISHING

ISBN 978-1-4671-2180-4

Published by Arcadia Publishing
Charleston, South Carolina

Printed in the United States of America

Library of Congress Control Number: 2013952062

For all general information, please contact Arcadia Publishing:
Telephone 843-853-2070
Fax 843-853-0044
E-mail sales@arcadiapublishing.com
For customer service and orders:
Toll-Free 1-888-313-2665

Visit us on the Internet at www.arcadiapublishing.com

To Buddy Thompson and all those that preserve our history for all to see

Contents

ACKNOWLEDGMENTS

I would like to thank first and foremost my wife, Laura, for digging deep to find patience over the years it has taken me to write this book. Managing a full-time job, a very busy (now pregnant) wife, a young child, sickness, and just life in general have been constant challenges for me in keeping this project going. Without Laura's support, this would not have been possible.

This book would also not be possible if there were not people that actually took pictures and wrote down and preserved information. I was constantly amazed and saddened at how little history is actually saved. I kept hoping to find "golden pictures," the ones that show everything I was looking for—the first stations, the first trucks, all the trucks, just every major change that took place—but many times I had to settle for disappointment. Luckily, enough was out there to put a book together and tell this story for the first time.

A big thank you to the *Asheville Citizen-Times* newspaper; without you and your predecessors recording area events and taking photographs, we would have no documentation and understanding of exactly what did happen. To local historians Bob Terrell, Nan K. Chase, Lou Harshaw, J.L. Mashburn, Nell Pickens, and Bill Alexander, for writing such wonderful local histories to draw from. To all those that opened their doors, called me back, and shared their archives and information with me. To local special collections at the University of North Carolina at Asheville, the North Carolina Room at Pack Memorial Library, Ann at the Swannanoa Valley Museum, and the Charles George VA Medical Center. To Biltmore Forest and the history they shared. And finally, to the Sanborn Map Company that documented our community in illustrated fire insurance maps to see: "just where was that?"

Thank you to everyone that gave me encouragement to keep going and their excitement about having this book.

INTRODUCTION

The Buncombe County of today was whittled down from an area that originally covered 12 North Carolina counties. It was started on April 16, 1792, at Gum Spring in Col. William Davidson's barn, just a few hundred feet past Biltmore Estate's Lodge Gate on the left. The county seat was set in Asheville, which was called Morristown at the time. The first charter was issued in 1797, and in 1883, Asheville became a city.

Firefighting in the early days was very simple and not so successful. With people so spread out and no waterworks in existence for a ready-made supply, organized fire departments were not possible. If people did have a fire, they would send family running on foot or on horseback to fetch the hopefully willing-to-help neighbors. Buckets, or anything that would hold liquid, were used to grab water from a nearby source and then dumped one at a time on the blaze. Wet blankets were also used to cover windows of any nearby buildings to help prevent the fire from spreading. If the fire was not caught in the early stages and one did not have an enormous amount of help, chances of success were low.

As populations grew and people began to settle closer together, it was possible to have enough people to form a dedicated fire department. The first apparatus used were hand-drawn carts or reels. They were moved on two or four wheels in different ways by two people on a wagon handle, by people all around pushing and a couple of individuals steering, or by people positioned like a bobsled team. Equipment was simple: single ladders, leather buckets, hooks used to pull down building material, and soda-acid chemical extinguishers, which used a chemical reaction to propel a fire suppressant.

When gravity-fed waterworks were installed with fire hydrants, it was possible to form hose companies that would simply connect 2.5-inch-diameter hose to a same-size discharge on the hydrant, put a diminishing-diameter, smooth-bore nozzle on the other end, and use a wrench to turn on the water from the hydrant. Apparatus used were carts with folded hoses or reels where the hose was wound around a wheel. Apparatus changed over the years, making use of horses and finally becoming motorized. Asheville and Biltmore were early departments that used these methods.

Protective clothing consisted of whatever one had on, which, unlike the clothes of today, was made of safer, natural materials. Eventually, rubber, wool-lined, long coats, along with boots and leather helmets, came around (more for protection from the water used on a fire than the from the heat).

The population of Buncombe County continued to grow and residents outside the city and towns needed help. The city answered the call to a point when they were spending too much time away from those they were formed to protect most. County residents needed a way to form their own community fire departments and pay for them. Another benefit would be the reduction in fire insurance rates.

North Carolina passed a law that allowed counties to contribute certain amounts of money to fire departments if they were properly supervised and met certain requirements. On September

28, 1953, the Buncombe County Board of Commissioners passed a resolution establishing a fire commission and rules for county fire departments to receive aid. George D. Young was appointed chairman of the commission and therefore fire commissioner for the county. The other two members of the commission were Chief J.C. Fitzgerald of the Asheville Fire Department and Chief E.M. Salley Jr. of the American Enka Corporation Fire Department. Col. P. Morton Keary of Skyland led the effort while establishing Skyland Volunteer Fire Department south of Asheville.

Approved fire departments received $50 per month, $100 at the end of the fiscal year in June, and a $1,000 loan repaid in 10 annual installments. Monthly payments were withheld if fire departments failed the following minimum requirements: roster of 10 men, fire chief, training program to include monthly drills, a traffic department with a captain and three members, identification for firemen, insurance, and maintaining certain firefighting equipment. There were 17 conditions in all. The departments were also required to be incorporated under North Carolina laws and meet minimum requirements of the state Fire Insurance Rating Bureau. Fire districts started out as a three-mile radius from the station. That distance could be increased to four miles if one met certain additional requirements, such as having at least a 1,000-gallon tanker truck, either superior means of alerting all members of the department or have station night duty, two self-contained breathing apparatuses (SCBAs), and a means for controlling traffic on scene. It was a big day when county departments passed their inspection by the Buncombe County Fire Commission and received their $1,000 check.

Residents began to meet in homes, churches, schools, and community centers to discuss plans and set the wheels in motion to form their own volunteer fire departments. American LaFrance (ALF) was present throughout the county and ended up selling many their first new fire truck. The standard one, which cost around $13,000, was on a Ford chassis, featured a 750-gallon-per-minute pump and a 500-gallon water tank, and was equipped with 1,000 feet of 2.5-inch hose, 150 feet of 1.5-inch, preconnected hose, a booster reel with one-inch hose, nozzles, and ladders. Many departments built their own first trucks and were filled with joy and pride when the first new fire truck arrived.

Several departments still have their first new fire trucks. They are Fairview, Black Mountain, Reynolds, Upper Hominy, Reems Creek, Barnardsville, Leicester, Woodfin, Jupiter, and Skyland.

When the volunteer fire departments formed, they relied heavily on industrious people to build the fire trucks, construct the stations, assemble all the equipment, and train. Several departments have members that have served for decades. None of this would be possible without the incredible fundraising efforts of the Ladies Auxiliary. Most fire departments had them and probably would not exist without them. They raised money through ham-and-bean, turkey-and-ham, hot dog, chicken, and covered-dish suppers; gospel singings; bluegrass concerts; square dances; horse shows; raffles; pecan, apron, candy, pie, rummage, flower, ice-cream, cookbook, and cake sales; gift shops; scrap drives; donkey ball; and basketball games.

They also provided needed services to the departments, such as bringing food and beverages to emergency scenes, answering telephones and running phone trees to alert the firemen of a call, keeping a yearly scrapbook of historical events, writing newsletters, providing annual meeting and Christmas dinners, holding food and toy drives for needy families and children, and even cleaning the station and washing the linens.

Most of the county's departments, even Asheville, had their own Ladies Auxiliary, and the Buncombe County Firefighters Association had one that met once a year. These groups were first made up mostly of firemen's wives but grew to include community ladies, as well as men (the first being at Reynolds in February 1987). Once fire taxes were started, their need diminished, and they began to dissolve. Fortunately, some still remain to support the fire departments.

Other agencies formed in the areas of dispatching, rescue, medical service, and forestry, all working with Buncombe County's fire departments to keep all who live in and enjoy this area safe.

One

The First Fire Departments

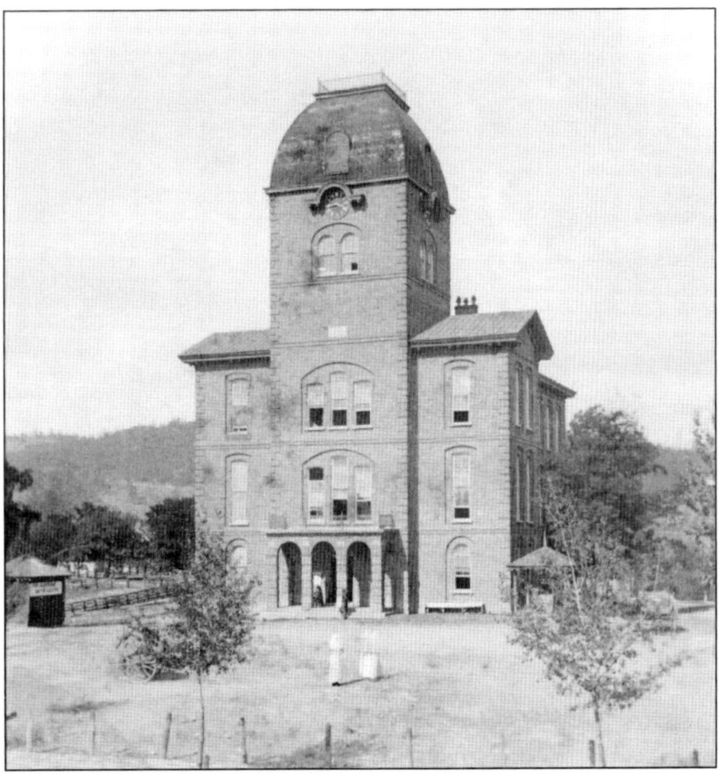

The Asheville Fire Department was formed on May 2, 1882. The first company was a hand-drawn hook and ladder, complete with 20 leather buckets. J.P. Sawyer was the first chief. On October 6, 1882, a motion passed to erect the first station made of rough plank next to the jail building behind this courthouse on the square. Headquarters were in the courthouse. The fire district was one-fourth of a square mile. (Courtesy of North Carolina Collection, Pack Memorial Public Library, Asheville.)

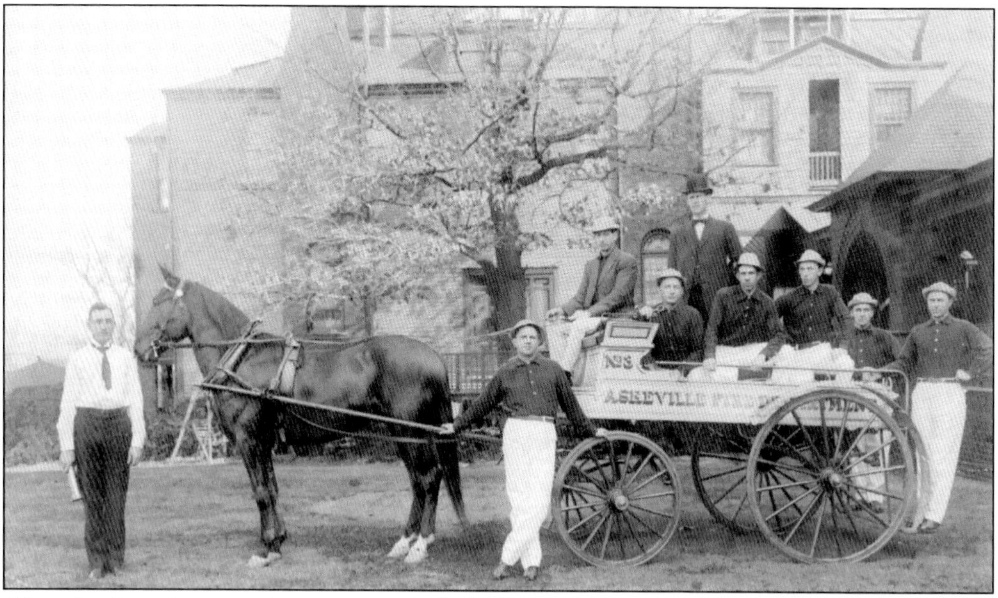

After the first city water lines were installed, Hose Company No. 1 was formed on August 7, 1884. The cart was hand drawn and held up to 1,500 feet of cotton hose, a hydrant wrench, and a nozzle. Firemen were usually worn out by the time they got to the scene. The first horse-drawn hose carriage was purchased in May 1887. By 1888, there were 40 volunteers and three hose carts. (Courtesy of Asheville Fire Department.)

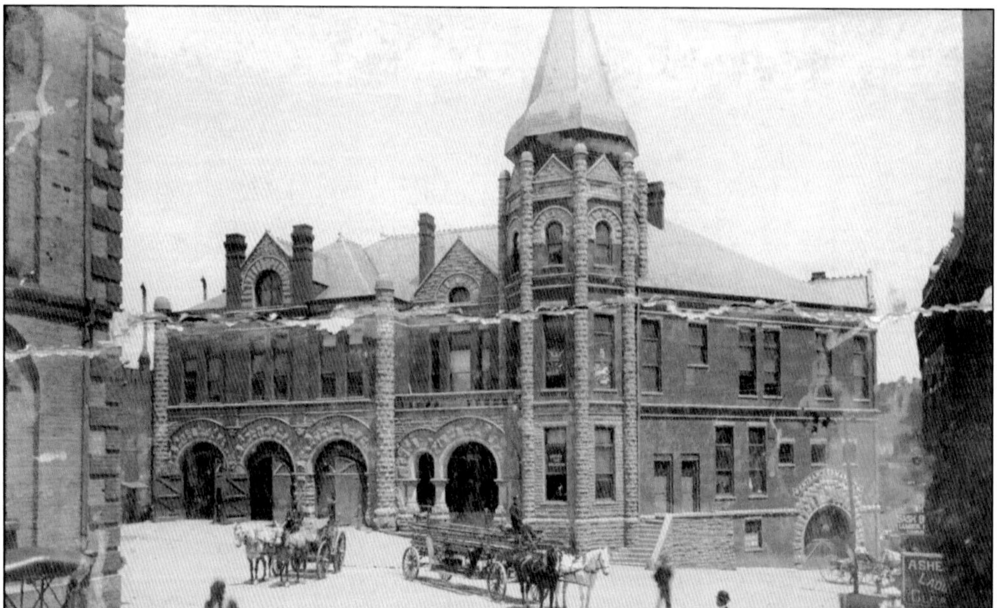

The first full-scale fire station opened August 2, 1892, behind the courthouse seen in the left corner of this image. The building included the fire department, police department, city hall, and a public market. A system of Gamewell pull stations was also installed around town, with a fire bell in the conical tower to alert the firemen. By this time, the fire department had four horses, two of which were named Dick and Tom, but still had hand-drawn apparatus. (Courtesy of North Carolina Collection, Pack Memorial Public Library, Asheville.)

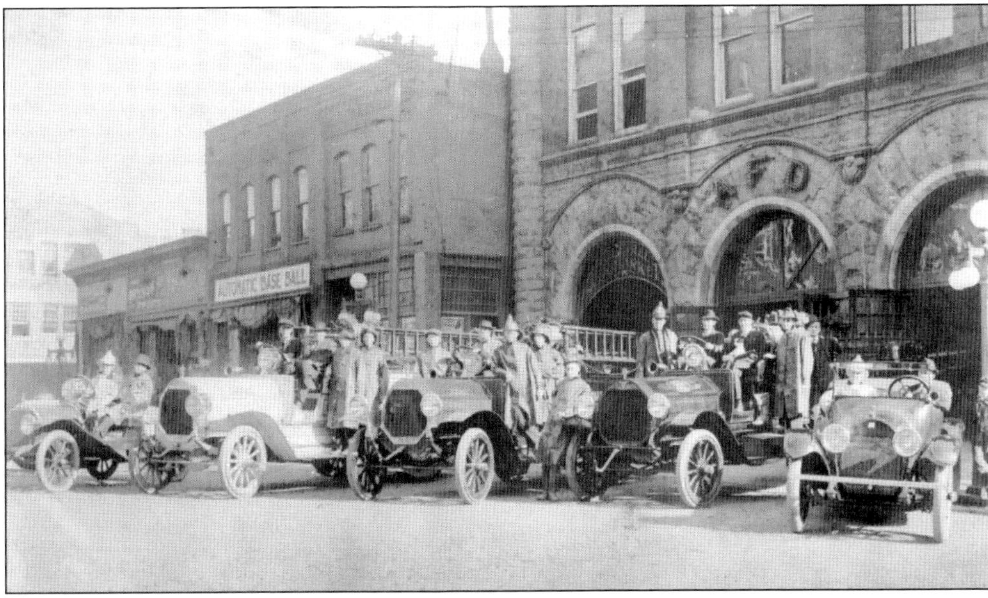

The first paid firefighters started by 1907 as horse-drawn drivers. The first motorized apparatus arrived in 1912, and by 1917 the Asheville Fire Department (AFD) was fully motorized. The first trucks were Seagraves. The six-cylinder, 80-horsepower engines had 45-gallon chemical tanks and 2,000 feet of cotton, rubber-lined hose. The hook and ladder truck included a 60-foot extension ladder, one life net, and 12 Hayward special respirators. (Courtesy of North Carolina Collection, Pack Memorial Public Library, Asheville.)

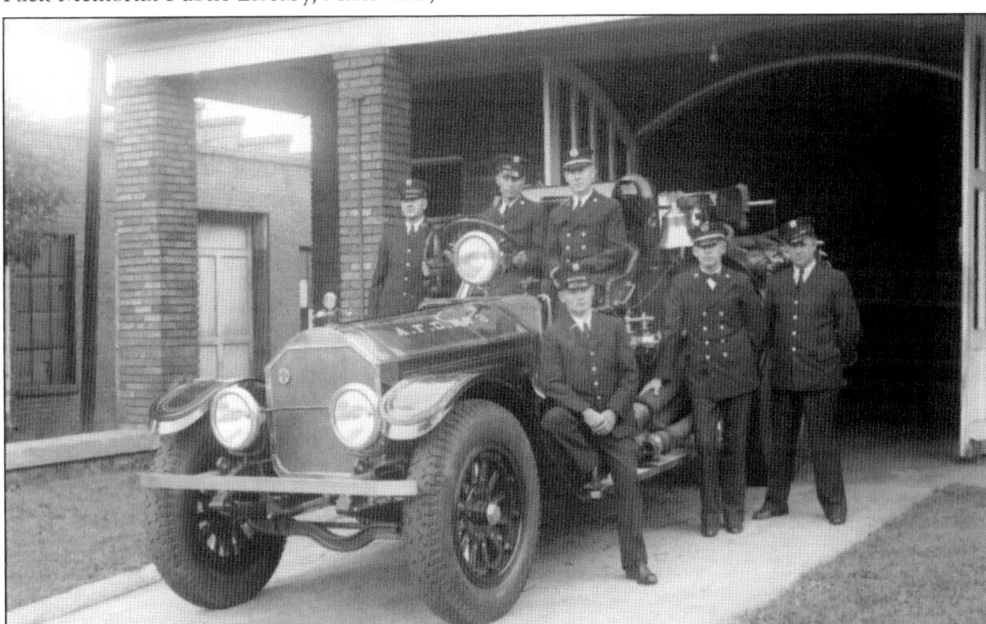

Station 2 opened on March 14, 1922, as Hose Company No.3. It was located on Bartlett Street to cover the industrial Depot section of town. The station included a single bay, a hose-drying room, an office, a kitchen, and even individual bunks. It was torn down during the urban renewal program after the current Station 2 opened in 1976. (Courtesy of Asheville Fire Department.)

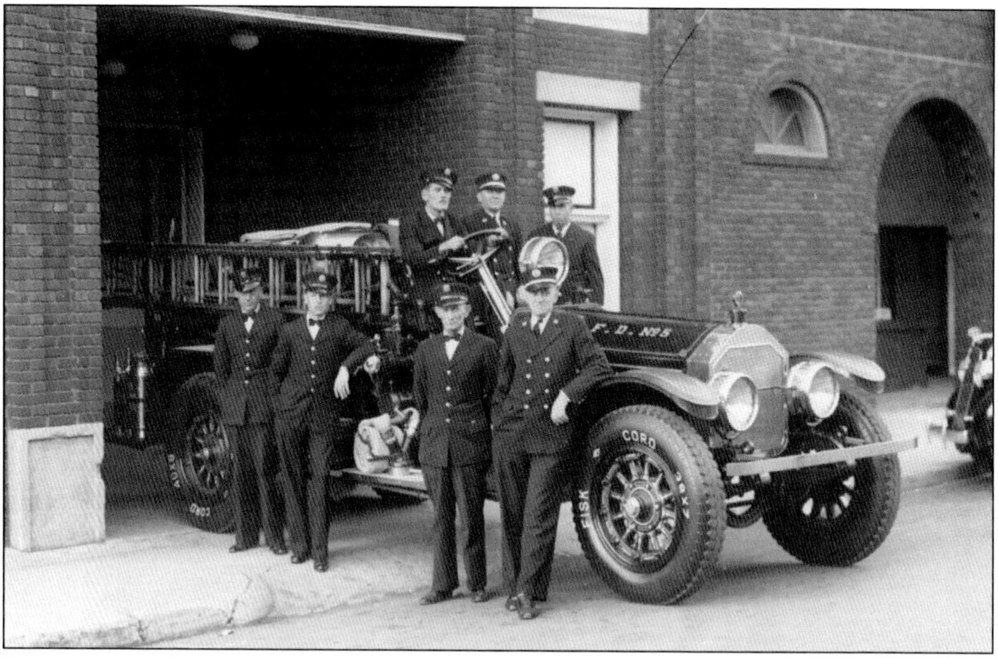

Station 3 started using the West Asheville Dunwell Avenue station when the city annexed the area in 1917. It was then closed for lack of manpower during World War I. A building two doors down from the above location was rented in 1918 for six men to sleep at night and use an REO truck. Station 3, above, opened in 1922 and was used until the current station opened in July 1979. (Courtesy of Asheville Fire Department.)

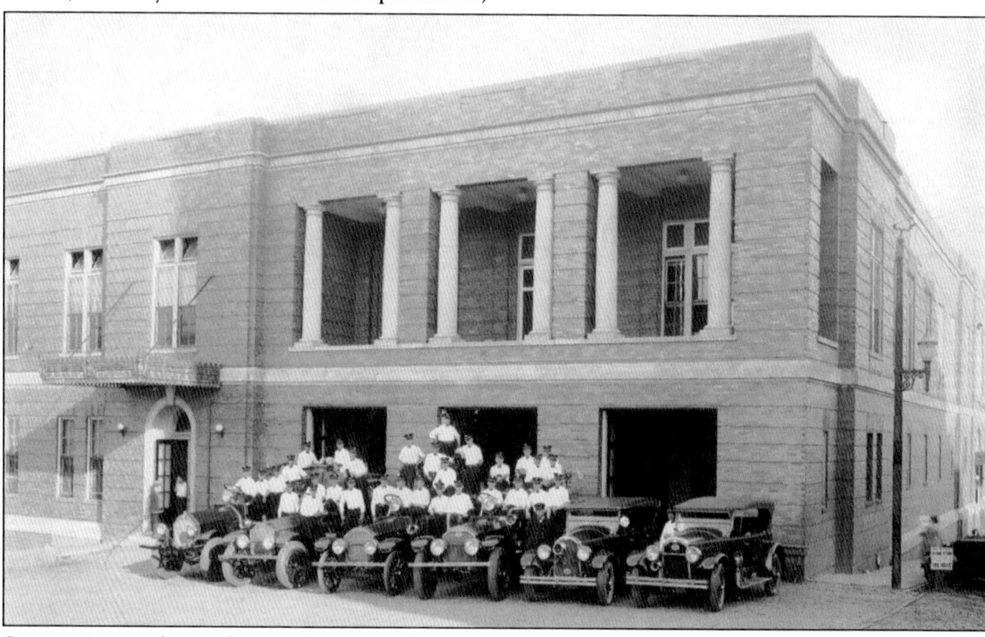

Station 1 moved next door to the original station starting March 8, 1926. This building was erected by former slave James Vester Miller, who ran one of Asheville's leading construction companies. It included the police department, a jail, a gym, a court of law, three fire poles, and a public market. The original station was torn down the same year. (Courtesy of Asheville Fire Department.)

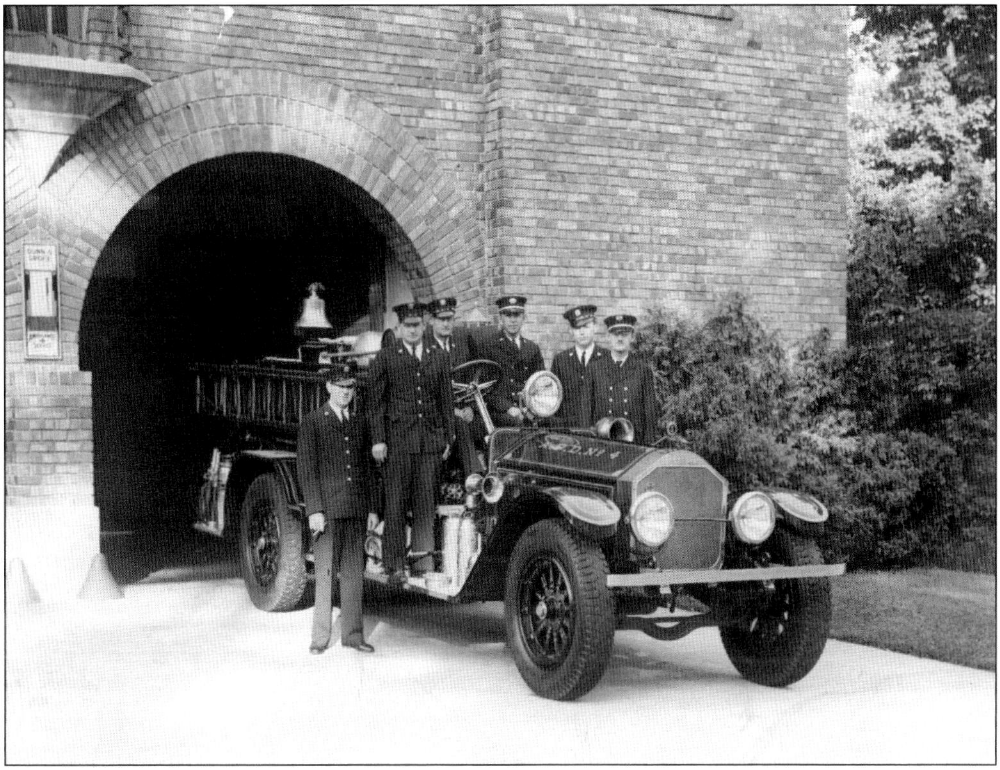

Station 4 opened on August 15, 1927. Famous Asheville architect Douglas Ellington designed the two-story station, which included two bays, a hose-drying room, a training tower, and four fire poles. It closed in 1976 and now houses the Asheville-Buncombe Arson Task Force. When the City of Asheville annexed Skyland's main station, it became Station 4. The current Station 4 houses AFD, Asheville Police Department, Buncombe County EMS, and Skyland Fire Department. (Courtesy of North Carolina Collection, Pack Memorial Public Library, Asheville.)

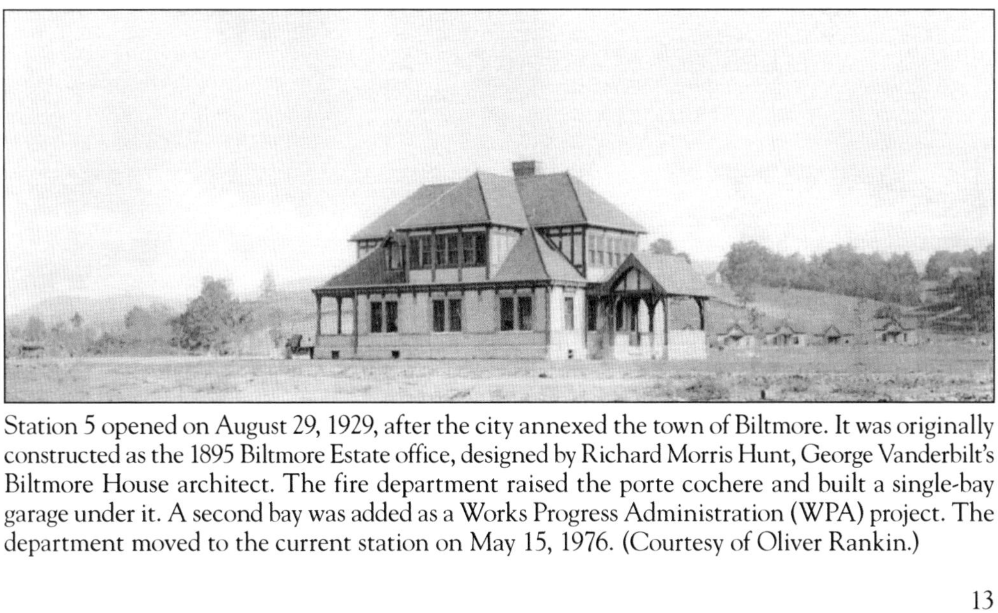

Station 5 opened on August 29, 1929, after the city annexed the town of Biltmore. It was originally constructed as the 1895 Biltmore Estate office, designed by Richard Morris Hunt, George Vanderbilt's Biltmore House architect. The fire department raised the porte cochere and built a single-bay garage under it. A second bay was added as a Works Progress Administration (WPA) project. The department moved to the current station on May 15, 1976. (Courtesy of Oliver Rankin.)

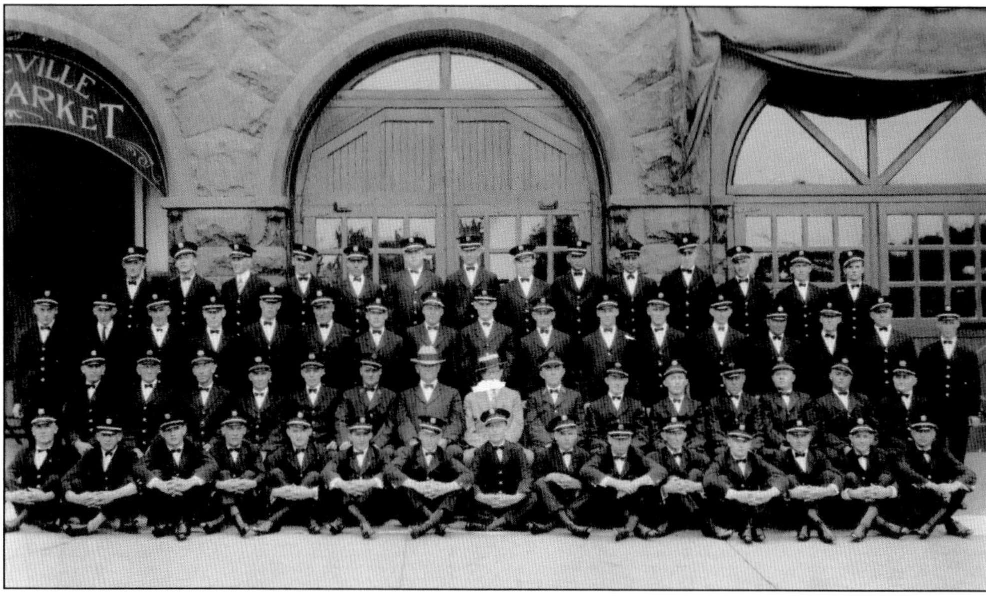

On April 1, 1924, AFD became a full-time, paid fire department of 54 firefighters. It would continue to grow, adding Station 6 in 1953, Station 7 in 1958, Stations 8 and 9 in 1960, Stations 10 and 11 in 2007, and Station 12 in 2009. The Gamewell pull-station system was disconnected in 1987. (Courtesy of Asheville Fire Department.)

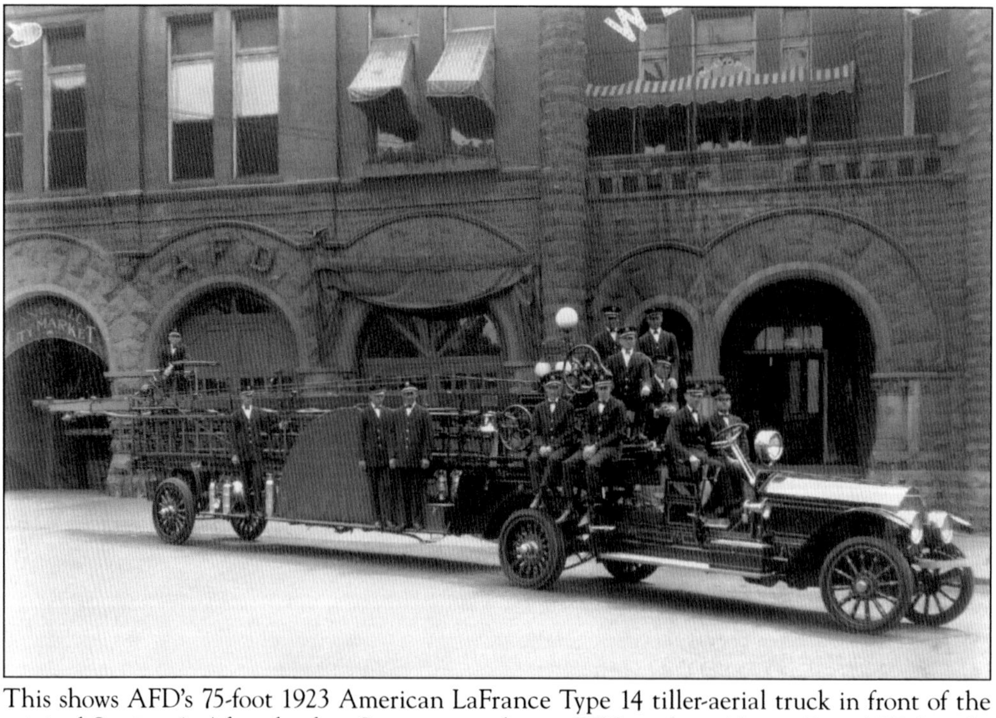

This shows AFD's 75-foot 1923 American LaFrance Type 14 tiller-aerial truck in front of the original Station 1. After the first Seagrave trucks, an REO and an Ahrens-Fox, AFD bought American LaFrance exclusively for many years. These trucks remained in service for a few decades. The last ALF was bought in 1984. (Courtesy of North Carolina Collection, Pack Memorial Public Library, Asheville.)

AFD firefighters killed in the line of duty include Gus Werhan (above, 23 Biltmore Avenue on December 9, 1963), Harry Williams (276 Haywood Road on October 20, 1944), Grady Wilson (Station 3 on June 28, 1948), Arthur Duckett (below, Station 5 on March 7, 1966), Jimmie Trent (the VA Medical Center on November 13, 1975), Raymond Flowers (926 Haywood Road on May 26, 1982), and Jeff Bowen (445 Biltmore Avenue on July 28, 2011). (Right, courtesy of Joan Werhan; below, courtesy of Bruce Duckett.)

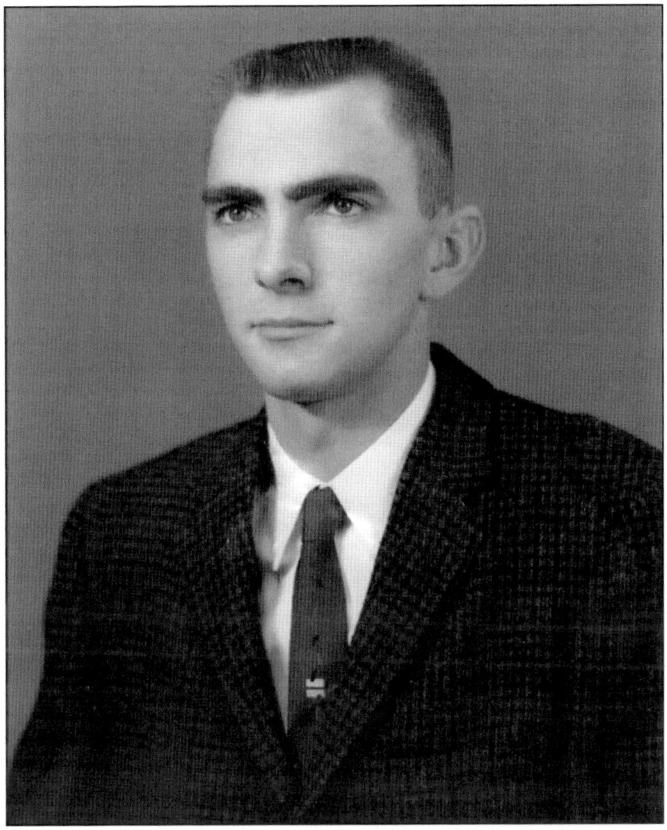

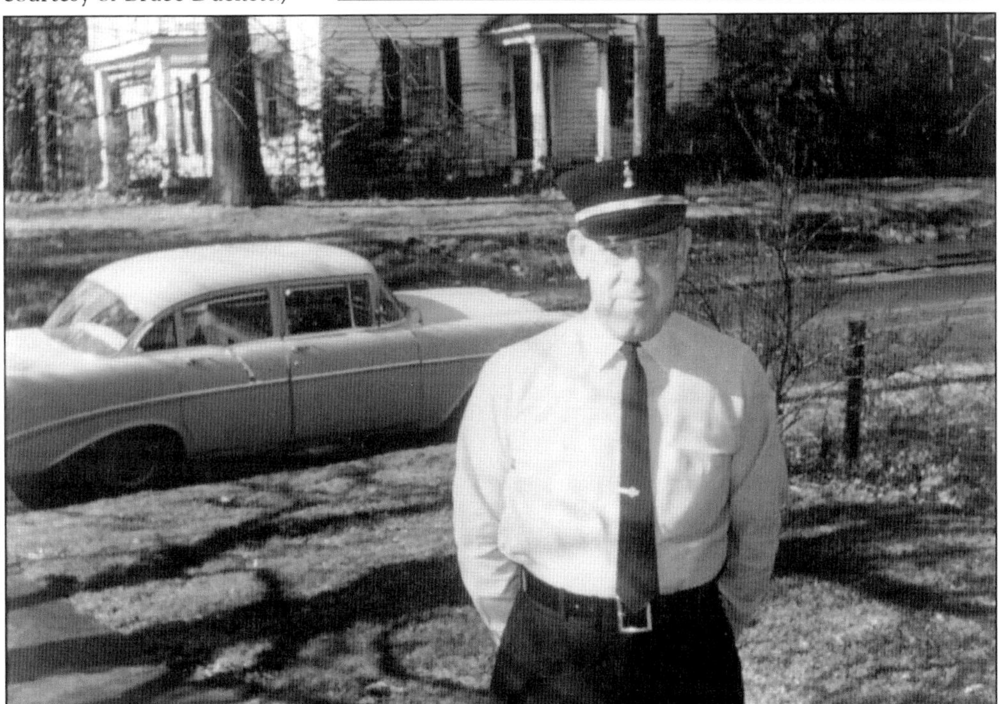

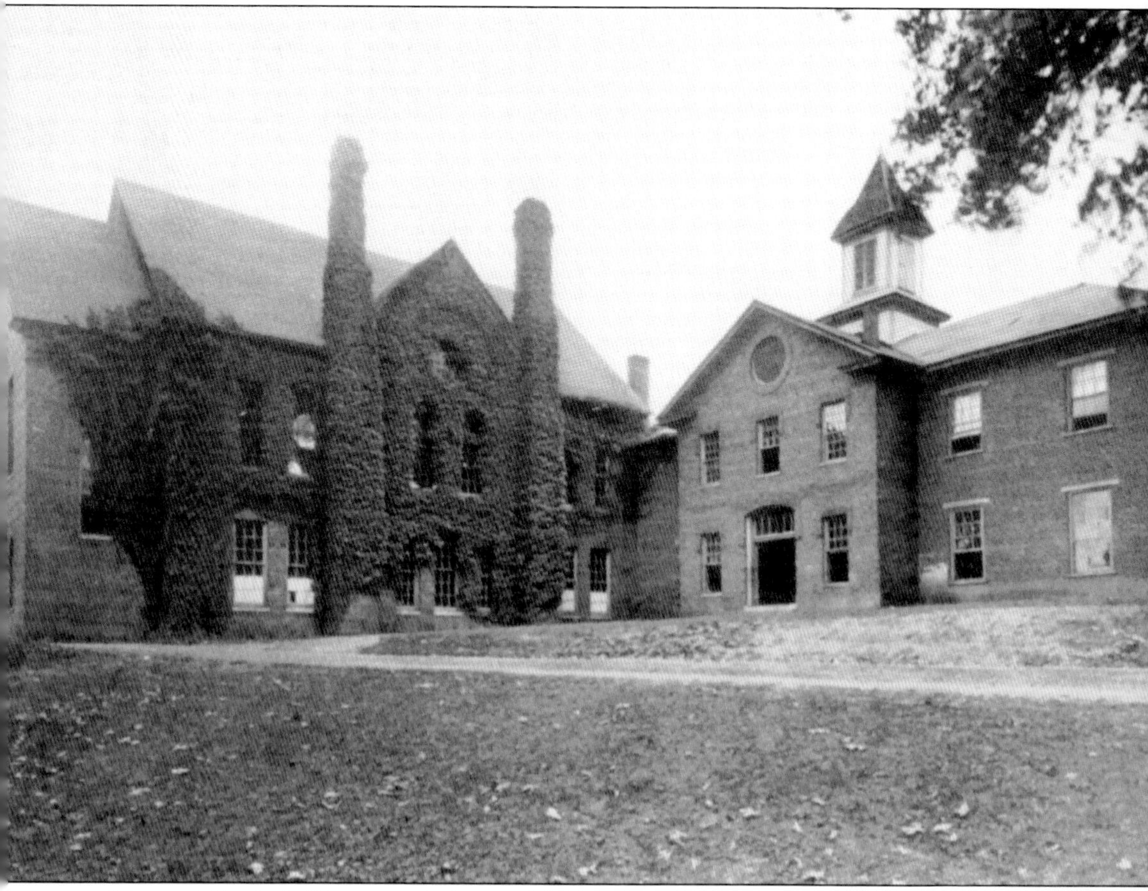

John and Elizabeth Weaver settled in an old Cherokee Indian area known as Dry Ridge sometime after 1786. A campground was established near the Weaver's home and was used for preaching, singing, and visiting; it became the incorporated Salem Camp Ground Association, and buildings were erected for church purposes. In 1873, Weaverville College was chartered near the campground to teach students from grade one through the college level. The college grew to over 55 acres and became the leading coeducational school in Western North Carolina. In 1934, it merged with Brevard College and Rutherford College, moving to Brevard and closing its Weaverville campus. The community's name became Weaverville in 1874 in honor of Rev. Montraville Weaver, who donated land and money to the Weaverville College. The name became official when the town of Weaverville was founded on March 16, 1875. Town limits were set half a mile from the J.W. Vandiver store. (Courtesy of Brevard College.)

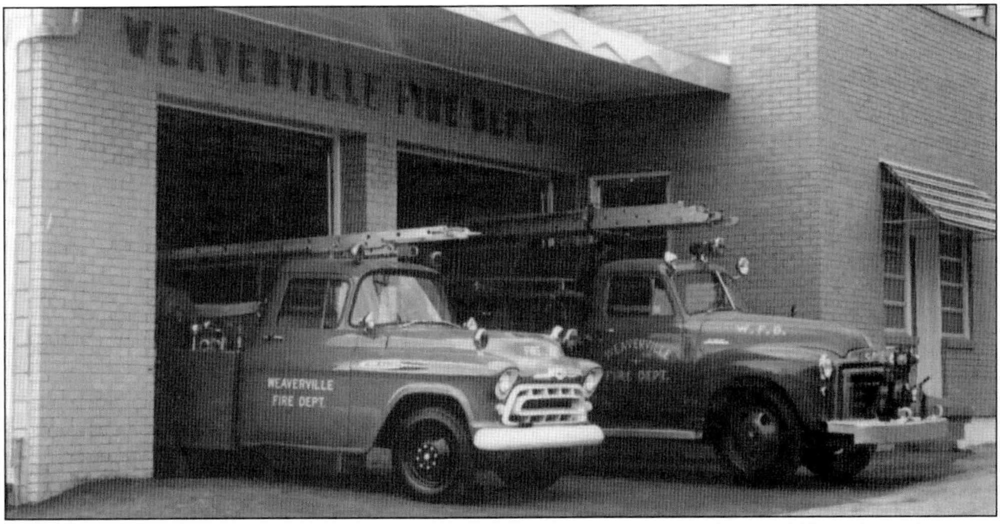

The first recorded mention of Weaverville Fire Department was on July 1, 1916, when the town hired Jim Murdock as fire chief. Little is known about the early years. Motorized fire trucks in those days included a former AFD engine and a 1927 Buick fire truck with wooden-spoke wheels, a wooden steering wheel, no cab, and no windshield. They were kept around town in different garages. In 1952, a GMC truck arrived, complete with a 500-gallon-per-minute pump and a 500-gallon tank. November 25, 1958, marked the opening of the two-bay station pictured above that was built by volunteers with donated materials and featured a meeting room, restrooms, and a kitchen. Weaverville, Skyland, and Barnardsville Fire Departments had rescue squads in 1961. In September 1967, they received a $16,000 Ford American LaFrance cab-over engine with a 750-gallon-per-minute pump and 500-gallon tank. This white truck, chosen because of cost, started their white color scheme. (Both, courtesy of Weaverville Fire Department.)

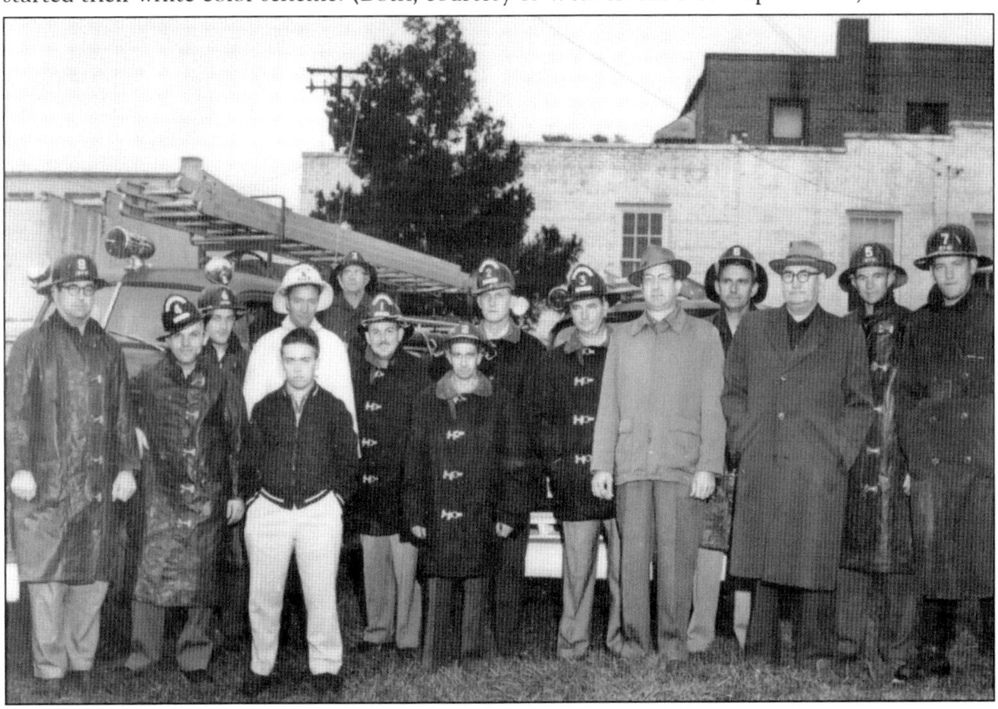

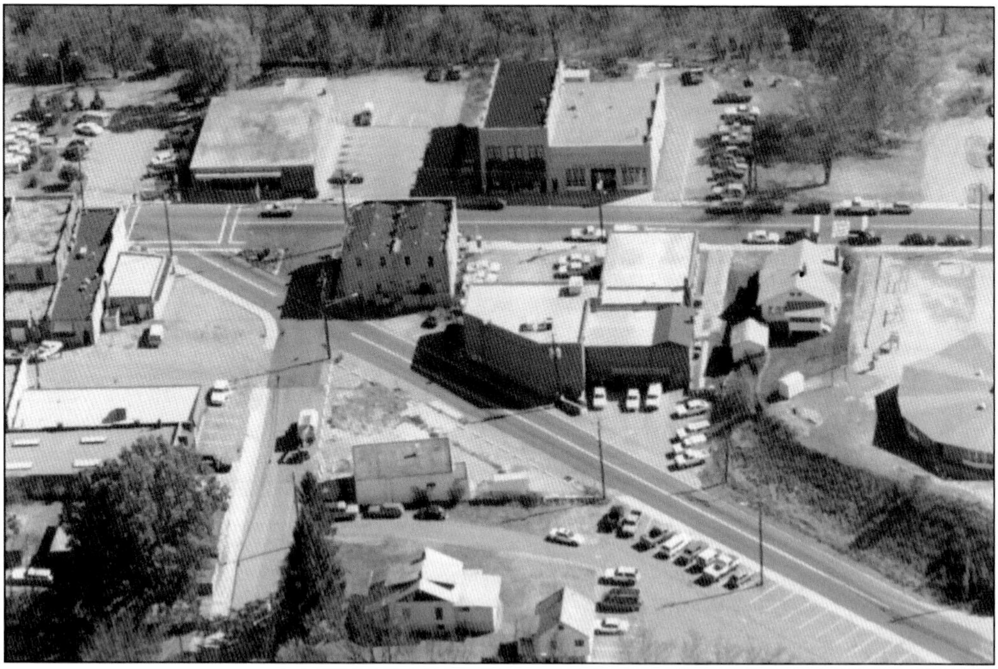

In 1980, Weaverville moved into the converted three-bay W.R. Reagan building on Main Street (shown at top center). Constructed in 1917, it had served as a general service store until the town bought it in the late 1960s. In April 1981, a plane carrying three federal prisoners made an emergency landing on Highway 25-70 and New Stock Road. Weaverville's first paid firefighter started in 1987. In November 1987, John Penley became the third generation in his family to serve as chief. In 1988, two bays were added on. In July 1989, the department received their first ladder truck—a Telesquirt Water Tower with a 1,250-gallon-per-minute pump and 500-gallon tank. During the aftermath of the Blizzard of 1993, from March 13 to 19, they responded to 711 calls. They moved into a new multimillion-dollar station in September 2007. (Both, courtesy of Weaverville Fire Department.)

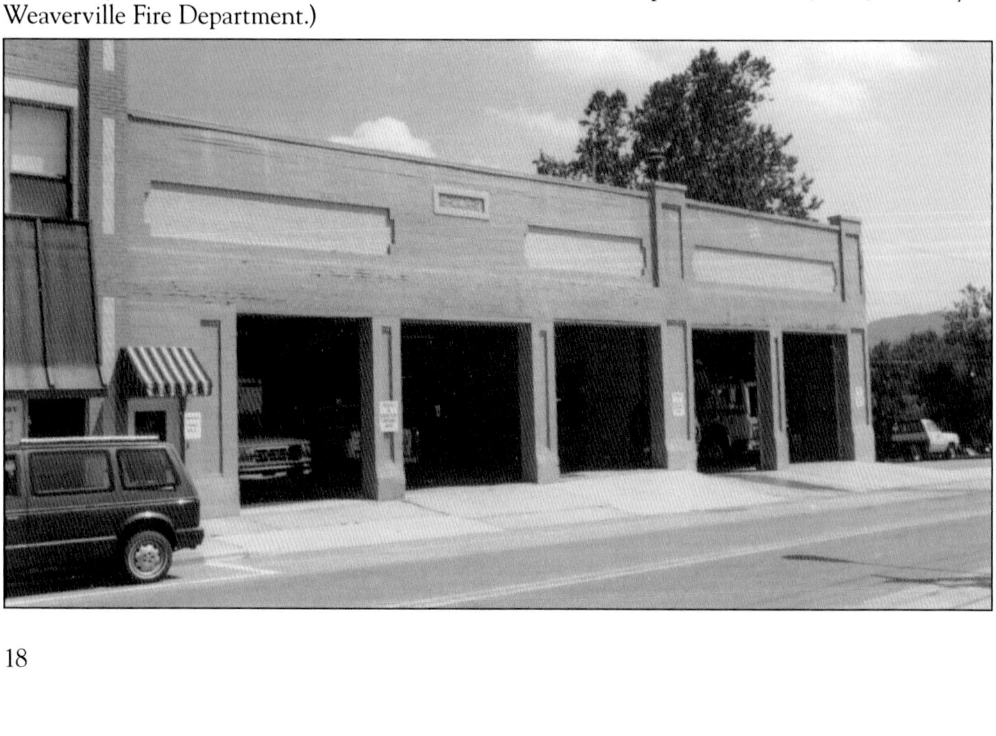

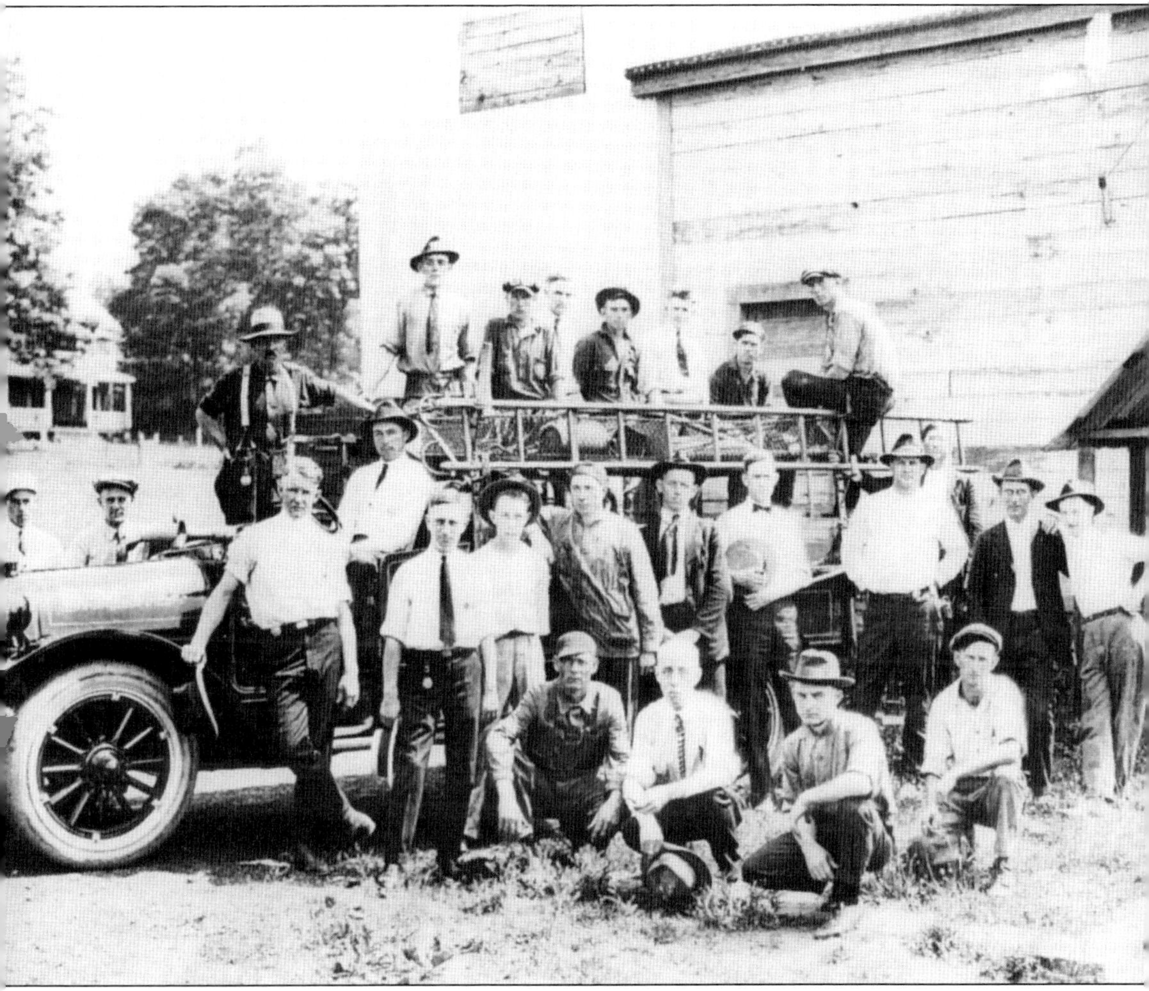

The town of Black Mountain was established in 1893. Several major fires occurred, including the burning of Black Mountain Hotel in 1906; Sen. Zebulon Vance's home, Gombroon, after 1915; and an entire downtown block on Cherry Street in 1912. The Black Mountain Fire Department was started on October 14, 1920, by 20 men. Their motto was, "I'll give." R.E. Currier was the first chief. They found scattered around town 800 feet of hose, two hand reels, a hand-drawn hook and ladder truck, and an old bell. The first station was in a vacant barn on Vance Avenue. A new facility was soon built on Broadway at the icehouse, shown above, and an alarm system was bought with nine stations in town and six outside of town. The first truck (above), an REO Speed Wagon, arrived on May 11, 1920. It was a commercial truck converted by the firemen. (Courtesy of Swannanoa Valley Museum.)

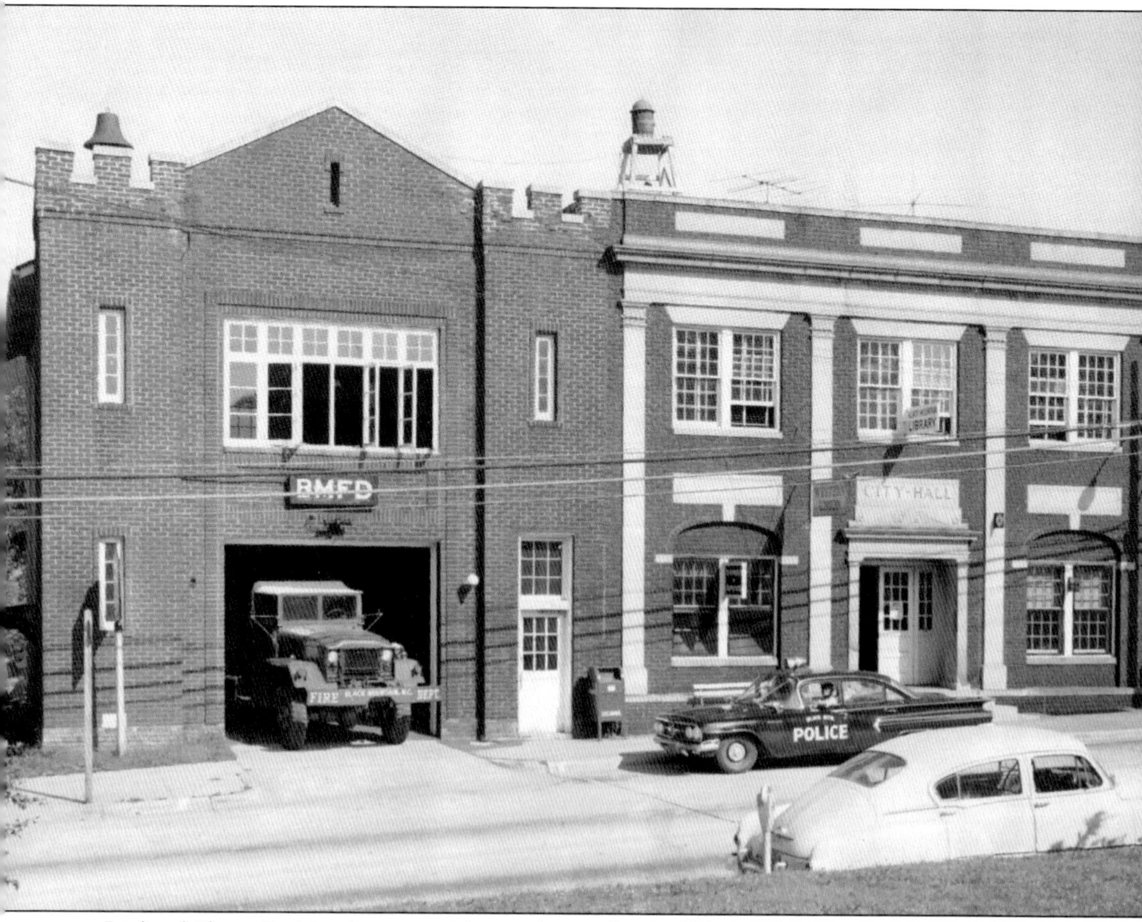

Richard Sharp Smith, the supervising architect for the Biltmore Estate, designed and built the above station in 1921. The station opened in 1922. In October 1922, the REO was traded for a former Asheville Fire Department 120-horsepower Seagrave combination truck. The department only paid $16 for it, and that was to have the name on the truck changed. City hall was added in 1927. (Courtesy of Black Mountain Fire Department.)

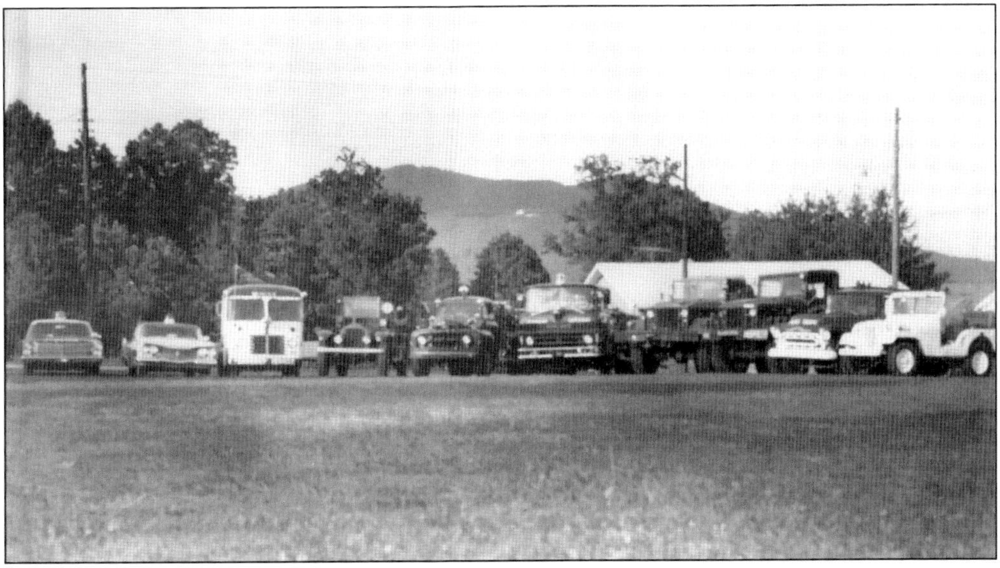

On May 15, 1926, a new 750-gallon-per-minute American LaFrance was bought for $12,500; it remained in service until 1964. Paid firefighters started as early as 1948. A rescue squad was started in 1962, and a cab-over Chevrolet American LaFrance was delivered on September 22, 1964. A five-ton Army surplus truck was bought and converted into a 1,500-2,000–gallon tanker in the 1960s. (Courtesy of Black Mountain Fire Department.)

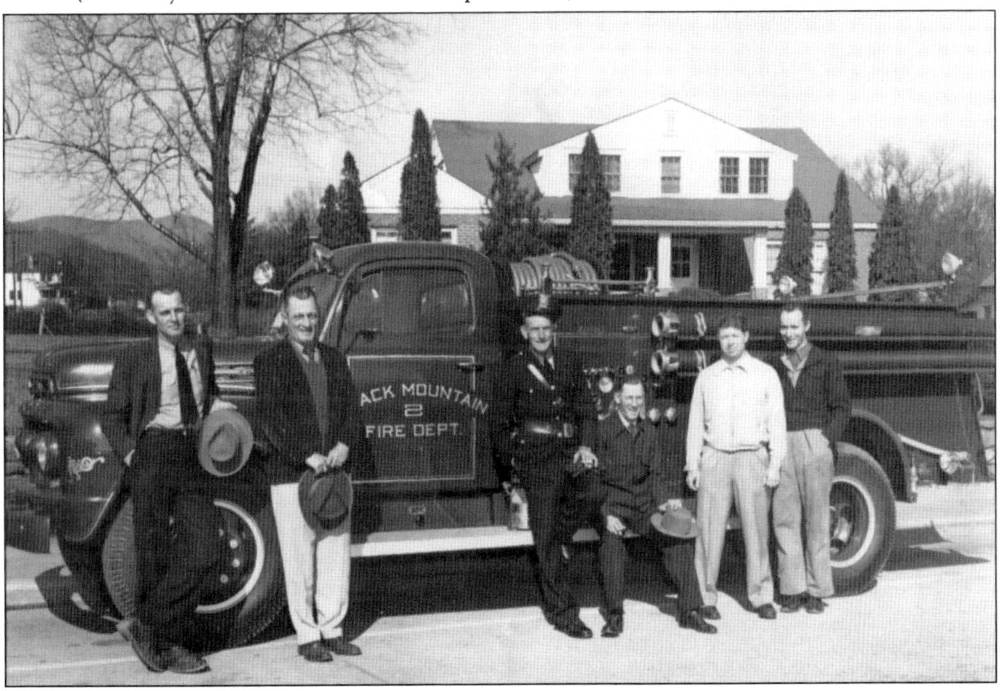

Black Mountain's current station was dedicated on August 30, 1987. Station 2 opened in 2003, and Station 3 opened in 2007. The fire district includes the East Buncombe Fire District and the town of Montreat, which, formed by the Mountain Retreat Association in 1897 for Christian work and living, included the Montreat College (opened in 1916). (Courtesy of Black Mountain Fire Department.)

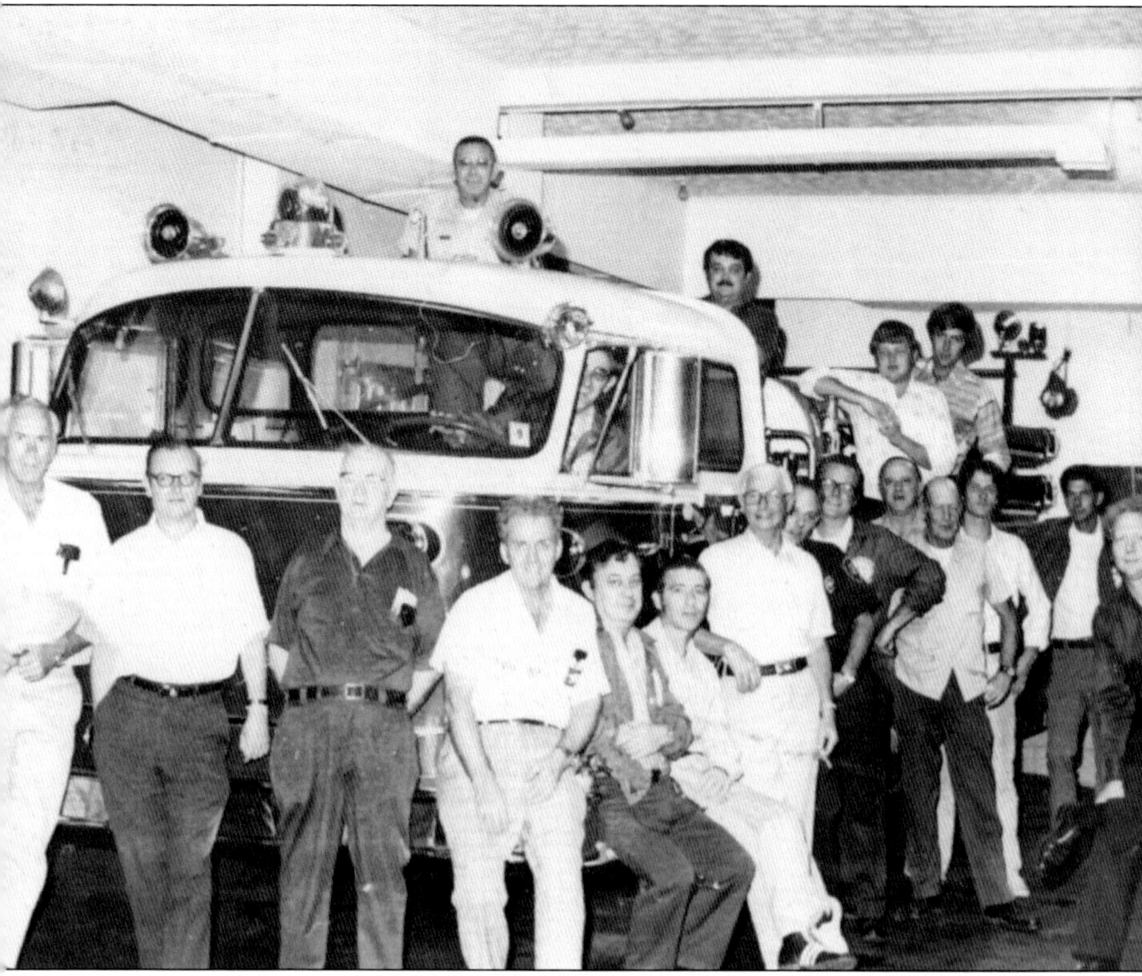

Charles E. Arnett retired in 1981 but went on to serve the fire department for a total of 72 years before he died at the age of 94. He served when the 1926 was the only one and had to be pushed downhill to start before driving to the fire. Here, he is seated at center, wearing a white shirt, on the corner bumper of an American LaFrance engine inside the 1921 station. (Courtesy of Black Mountain Fire Department.)

Two

Buncombe County Volunteer Fire Departments

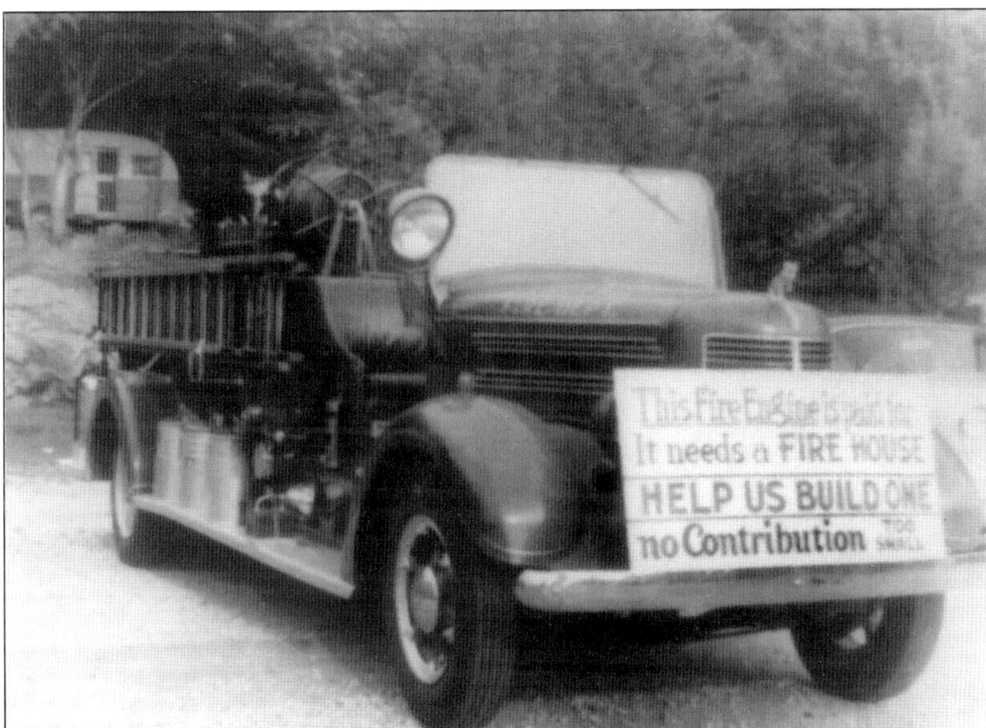

On February 20, 1953, a meeting was held at Valley Springs to discuss the formation of the Skyland Volunteer Fire Department. A study committee was formed. This started after Ken Cunningham's home burned in Rosscraggon. Col. P. Morton Keary, real estate operator of Royal Pines, had been guiding the movement. A mass meeting was held on October 29 to discuss the results of the study with the Limestone Township and surrounding communities. (Courtesy of Skyland Fire Department.)

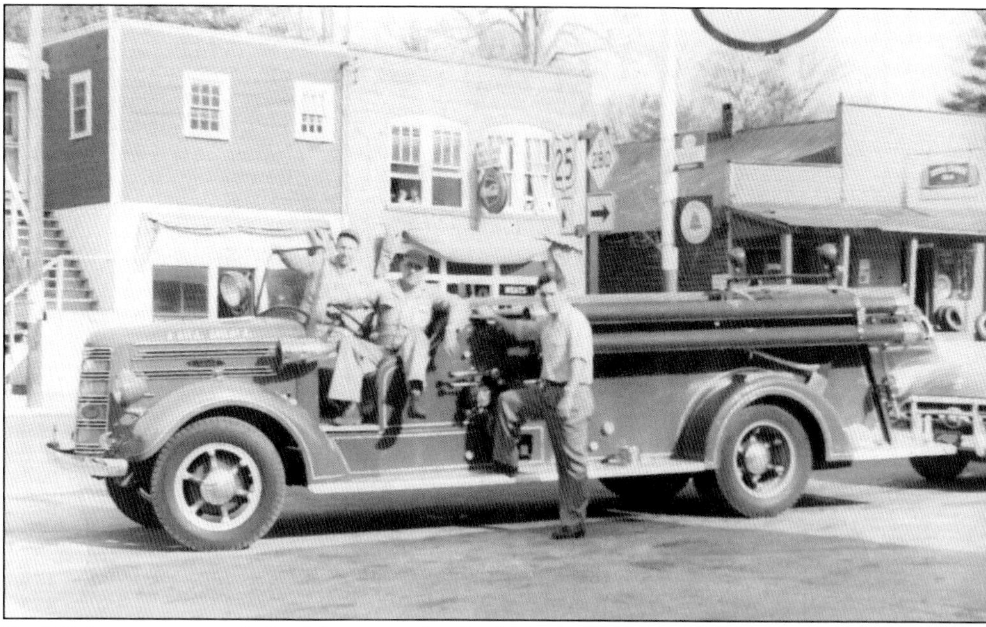

Fred Dalton was named chief. Property for the firehouse was located next to the post office (to the left with the stairs). The first truck, a 1941 Mack 500-gallon-per-minute pump, was purchased from the city of Greenville, South Carolina's Parker Water District. It arrived in the beginning of April 1954 and was first kept at Fred Dalton's Esso service station. A charter was issued in November. (Courtesy of Skyland Fire Department.)

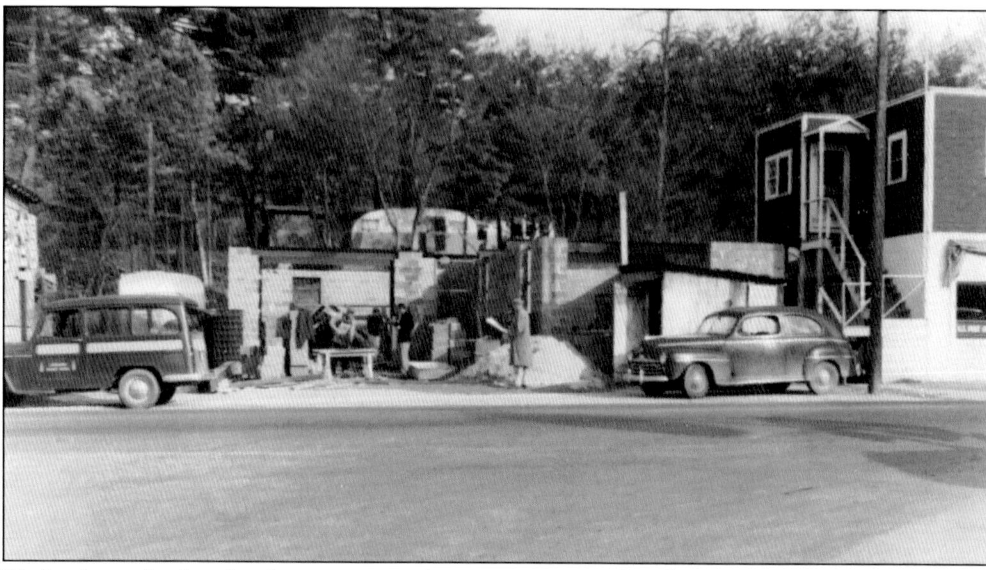

Skyland was the first fire department in the county set up after the new Buncombe County Fire Commission formed under a state law, which permitted counties to contribute certain amounts for fire department maintenance when properly supervised and meeting certain conditions. Ground was broken on the new one-bay, two-story, $8,000 station in October 1954. A meeting hall upstairs provided space for dances in which the Fireball Five, a string band made up of local firemen, could raise money. The first floor was completed in time for the April annual meeting. (Courtesy of Skyland Fire Department.)

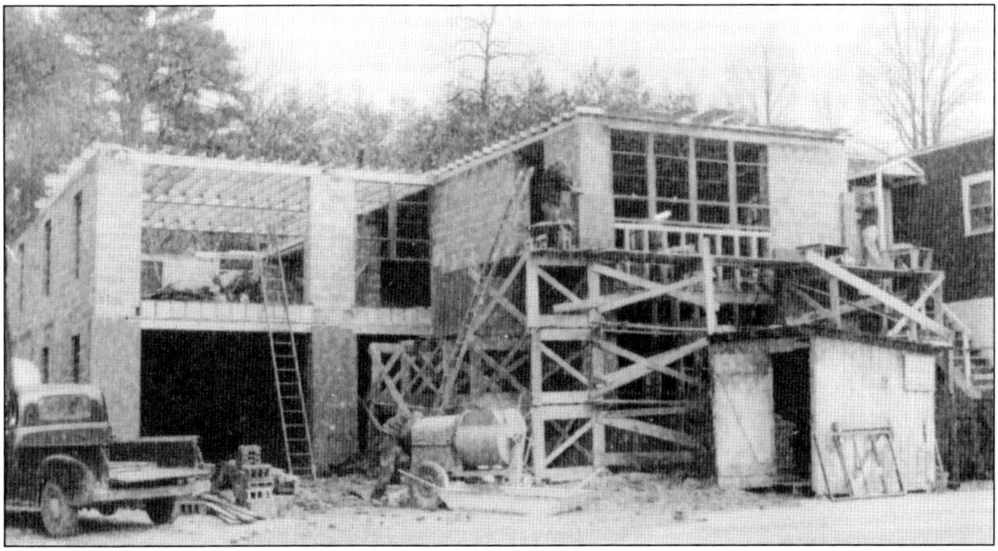

Alarms were first called in to the chief's Royal Pines home or his service station. A firefighter was on duty all night at the station. The Skyland Post Office rented an 800-square-foot space on the first floor starting in March 1955, moving from a building constructed about 17 years prior. Before that, it had been in a building housing the Skyland Soda Shop. The post office started in 1889 on the south corner of Long Shoals and Hendersonville Roads when Skyland was a tourist destination complete with a hotel and train depot. (Courtesy of Skyland Fire Department.)

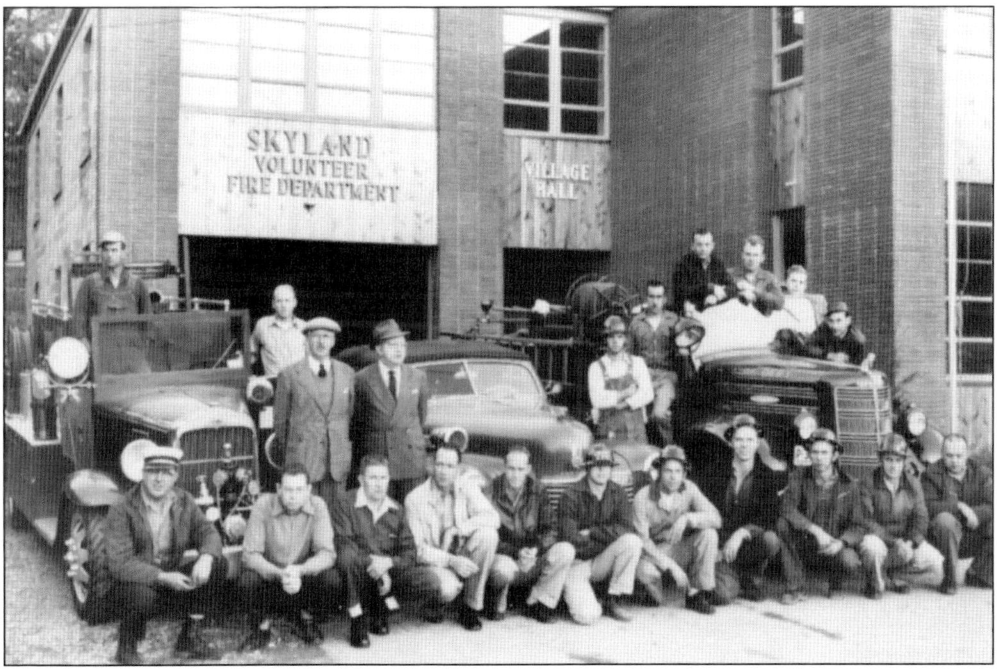

In November 1955, Wesley Kaupp donated Skyland's second fire truck from the Edison Electric Company in Detroit, Michigan. Seen above at left, it was a fully equipped 1928 REO with an 850-gallon tank. The department's first rescue occurred in February 1956 when a rain-soaked cat was retrieved from a tree. Work on finishing the second story was started in January 1956 and completed in 1957; the space was held for a village hall. (Courtesy of Skyland Fire Department.)

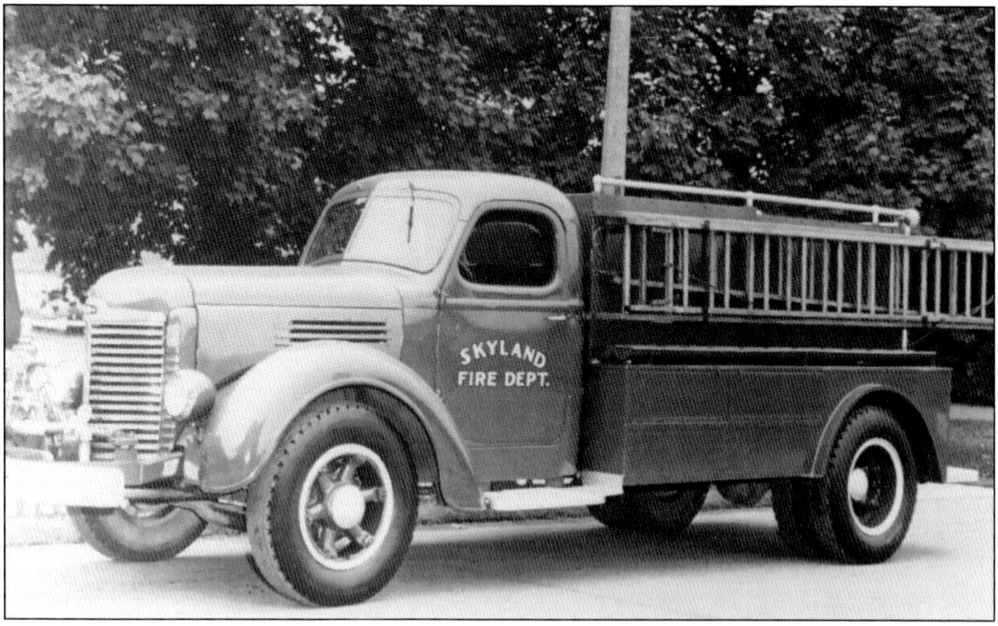

The first-annual Fire Department Day was started August 1957 as a way to look back at the year's accomplishments. The body of the REO was later moved to an Army surplus chassis in 1958, as pictured above. The first wedding at the station occurred July 1958. During severe weather in February 1958, firemen remained on-duty for 24 hours to keep the trucks from freezing since the heating system was not installed until 1959. (Courtesy of Skyland Fire Department.)

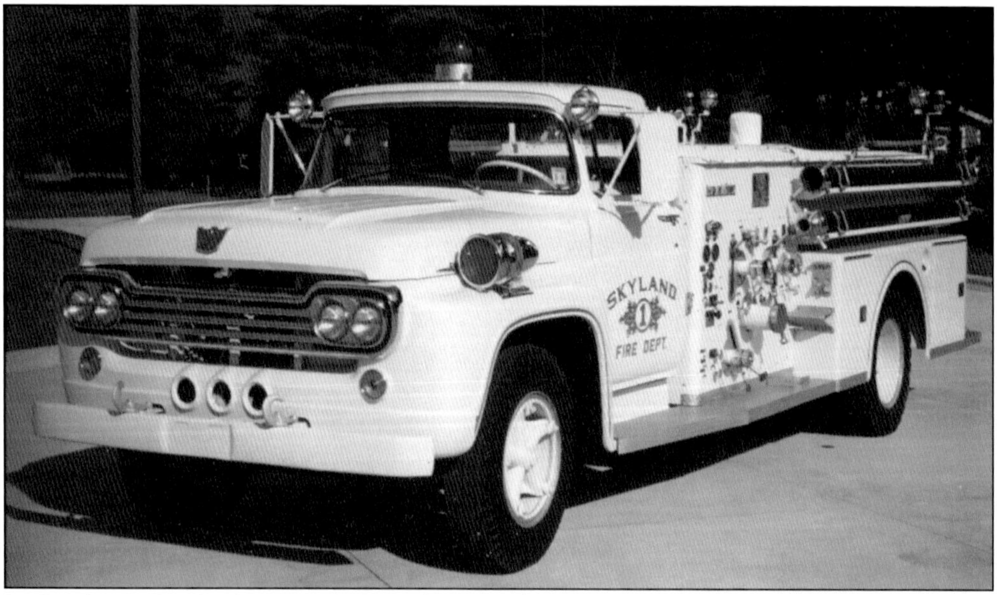

In October 1960, the first new ALF Ford, 750-gallon-per-minute, 500-gallon tank arrived. The Avery Creek substation opened on October 26, 1960. It was a single-bay, concrete-block station built by members of the Avery Creek Community Club. The Mack truck was its first truck. In 1963, Buncombe County fire departments split their radio traffic from the sheriff's frequency and started mutual-aid agreements. Firemen also started repairing toys for children that year. (Courtesy of Skyland Fire Department.)

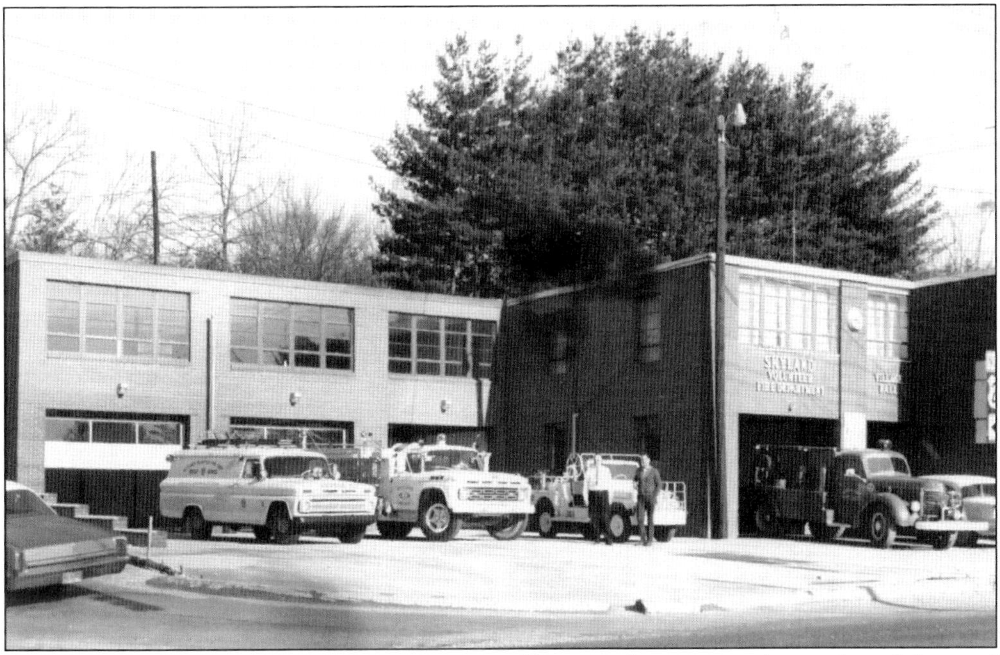

In June 1960, construction started on a two-story addition to the rear, complete with workshop and recreation room upstairs for the firemen. Costing $17,000, the above 750-gallon-per-minute truck with foam injector arrived on a Southern Railway freight car at the Biltmore depot in September 1964. In February 1966, the pictured three-bay third addition, built onto the rear second addition, was completed. (Courtesy of Skyland Fire Department.)

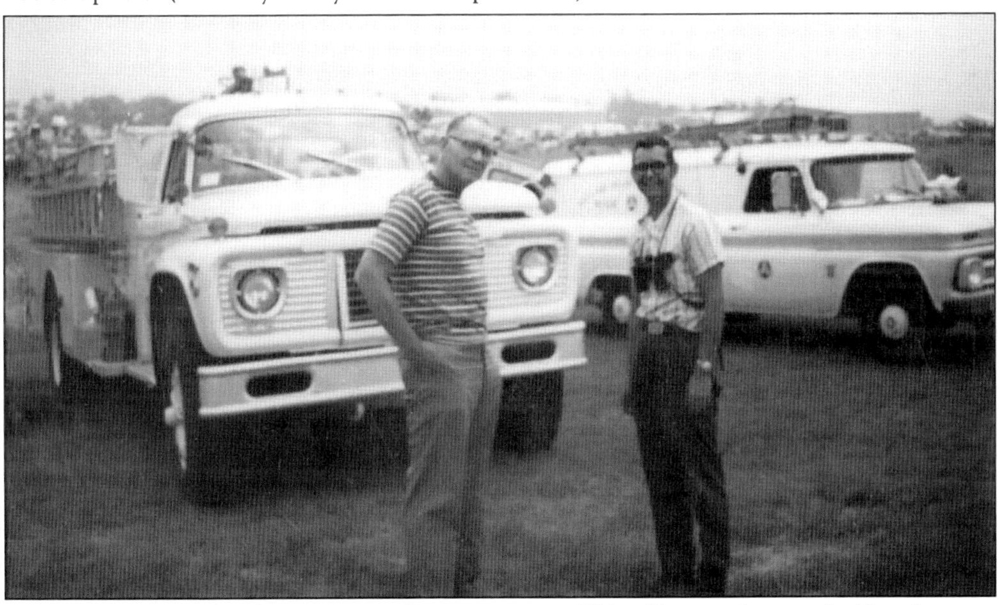

On April 8, 1968, the above 1,000-gallon-per-minute, 750-gallon truck arrived. Also shown is the 1963 Chevy rescue truck. In May 1973, a major flood hit the area, causing much destruction and death. A four-bay metal building was constructed by the end of 1973 behind the station, and the north-end bay was later turned to the side to form two bays. (Courtesy of Skyland Fire Department.)

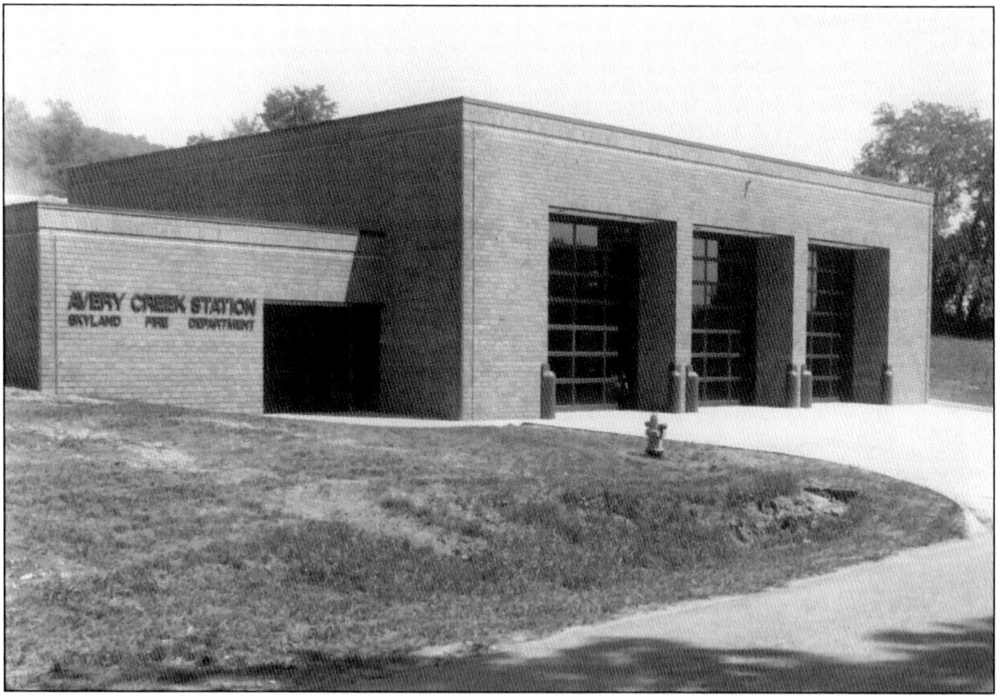

In 1978, the new Avery Creek substation—built in front of the original—was opened. A dedication was held on June 3, 1979. An addition was built on top of the left wing and behind the station in 1991. A six-bay addition, complete with individual bunks, locker room, day room, and a community center, was added on the rear in 1998. (Courtesy of Skyland Fire Department.)

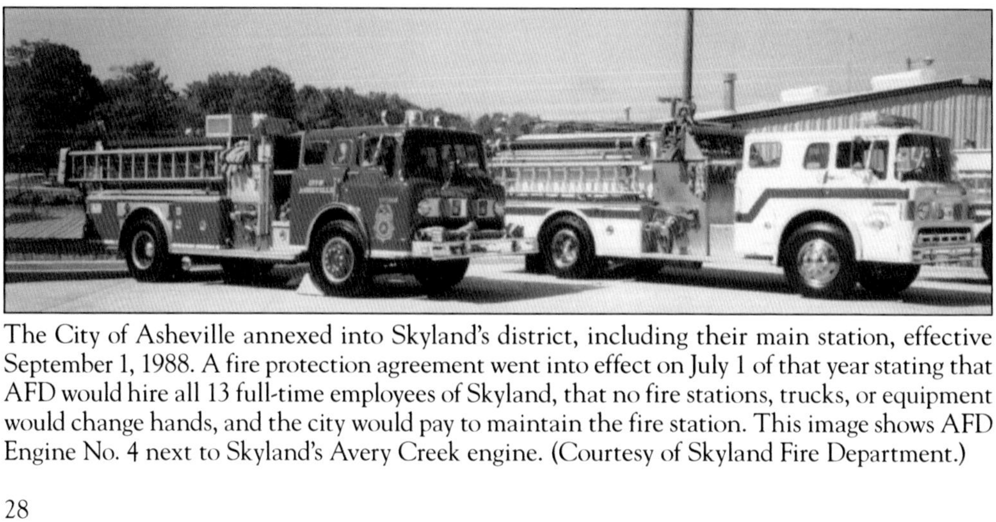

The City of Asheville annexed into Skyland's district, including their main station, effective September 1, 1988. A fire protection agreement went into effect on July 1 of that year stating that AFD would hire all 13 full-time employees of Skyland, that no fire stations, trucks, or equipment would change hands, and the city would pay to maintain the fire station. This image shows AFD Engine No. 4 next to Skyland's Avery Creek engine. (Courtesy of Skyland Fire Department.)

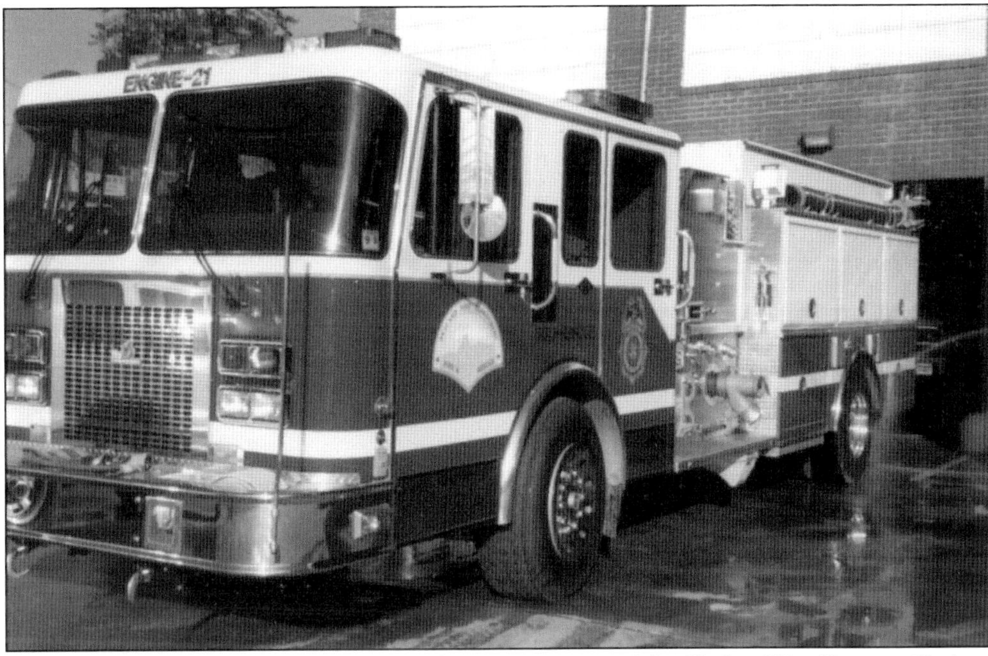

Additionally, AFD would provide day-to-day management of the Skyland Fire District, Skyland would retain control over fire services in its district, and volunteers could still serve outside the annexed area. The terms of the agreement have changed over the years. This photograph shows the only joint AFD-Skyland bought engine. (Courtesy of Skyland Fire Department.)

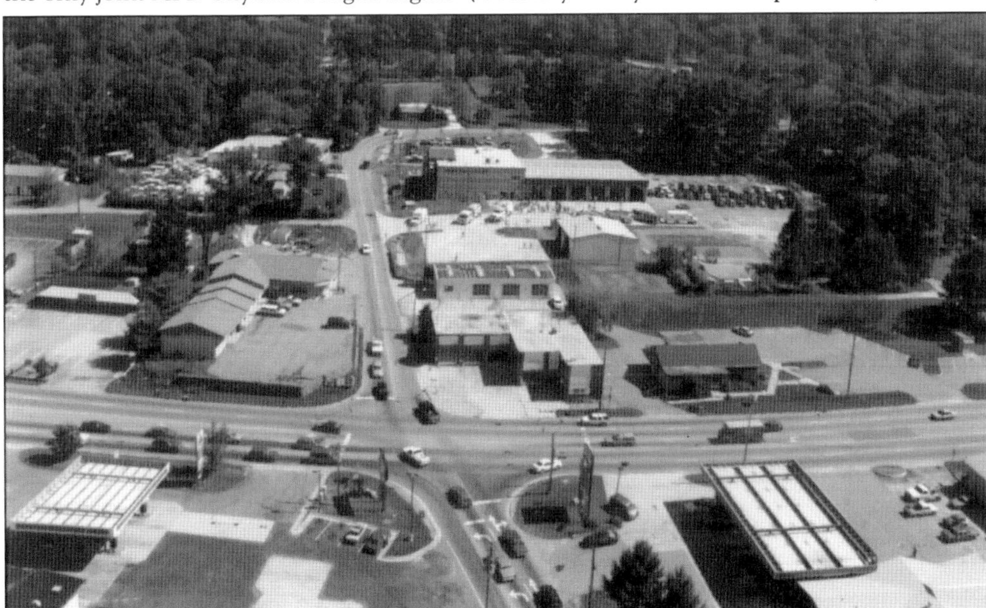

Skyland's third station, named for John E. Bishop (a longtime volunteer and chief), was dedicated September 11, 1994. The new 12-bay main station, above, was dedicated on October 5, 1997. The station included individual dorm rooms, a day room, a kitchen and dining area, a library, classrooms, a reception area, and a exercise room. AFD added on four bays, offices, and storage to the south end and celebrated with an open house in May 1999. (Courtesy of Skyland Fire Department.)

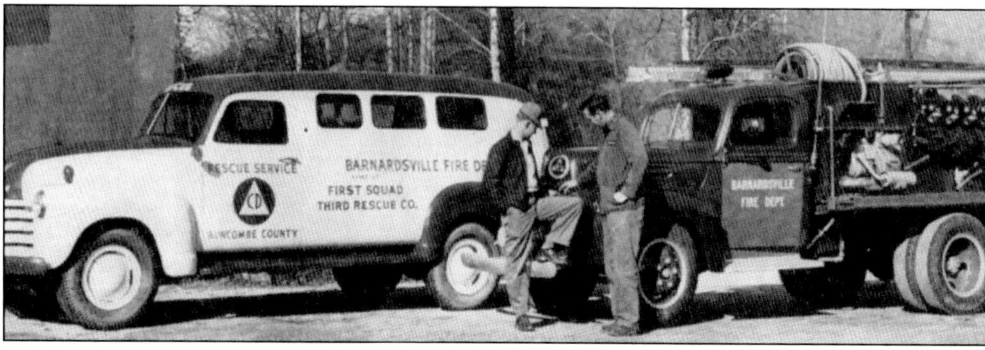

A meeting was held May 13, 1958, to organize the Barnardsville Fire Department. Dick Dillingham was named the first chief. First trucks included a 1939 Ford pumper. A former 1952 Chevy West Funeral Home ambulance was used to form a rescue squad in 1959. The pumper was bought by the Big Ivy Men's Club and had a front-mount pump that would just about jerk the operator's arm off. (Courtesy of Barnardsville Fire Department.)

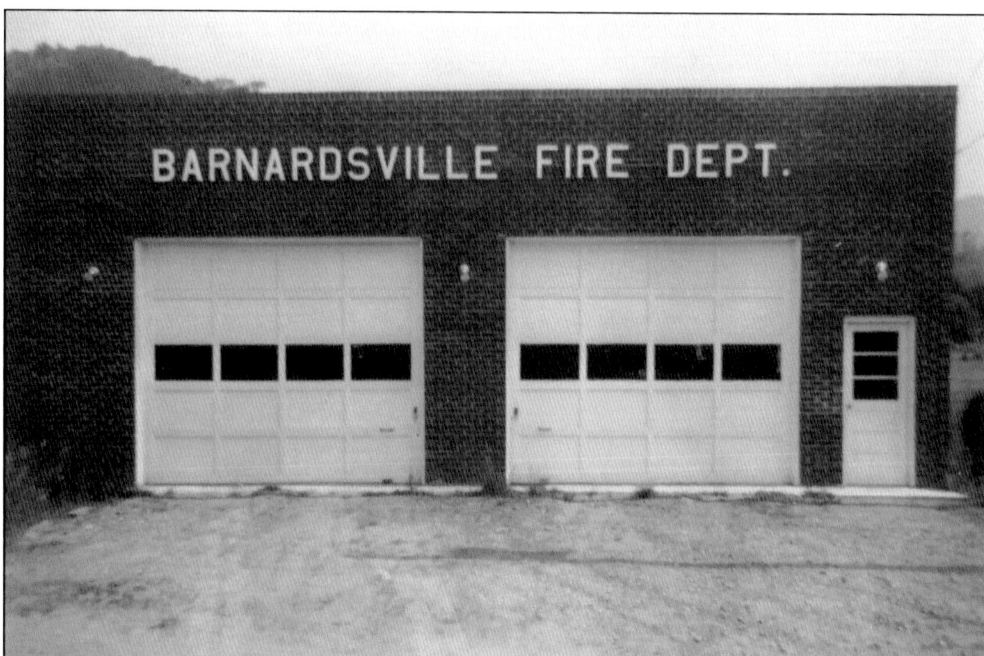

The department's first fire was at Mayor E.W. Cannon's house in May 1958. Despite the crew's lack of training, the fire was under control when Weaverville Fire Department arrived from 10 miles away. The first station was in a garage at Chief Dick Dillingham's store. Land was purchased for a station and town hall, and construction was completed in 1960. The fire truck was housed in Mayor Cannon's tobacco barn while this station was being built. (Courtesy of Barnardsville Fire Department.)

The station was not built until after the main road through town was relocated. Communities in the district included Dillingham, Democrat, and North Fork. Barnardsville was an incorporated town complete with fire and police service and a women's garment factory: Clinton Corp. of North Carolina. George Harris was the last mayor when the town dissolved in May 1965 for financial reasons. This September 1964 image shows original trucks, including the tanker at right. (Courtesy of Barnardsville Fire Department.)

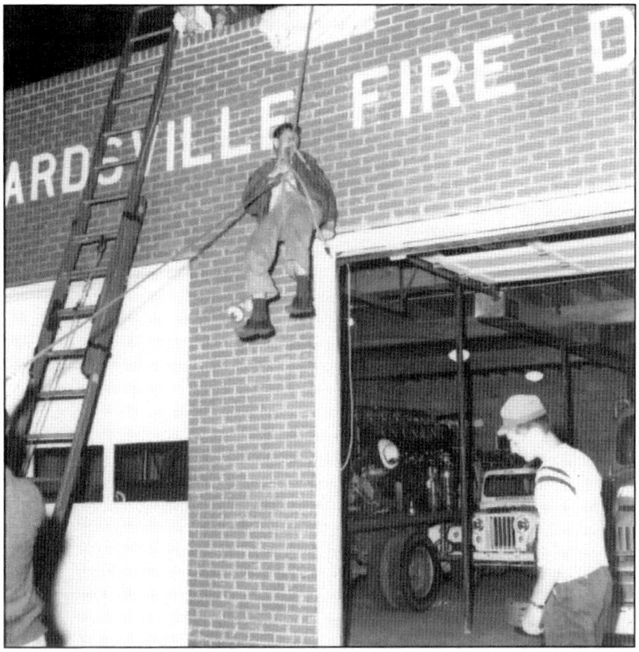

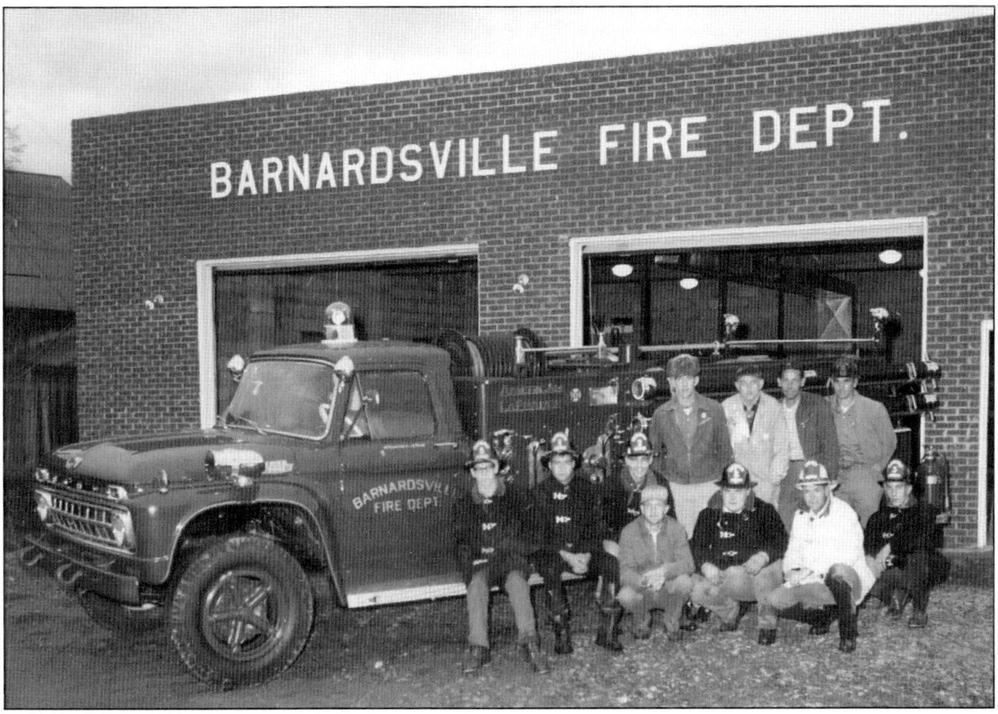

Barnardsville's first new truck, which is still with the fire department today, was received in December 1965. On January 4, 1973, a $20,000 loan was approved for a two-bay addition, complete with a meeting room and kitchen. Ambulance service started in 1974. In March 1981, the department had 30 firemen, 14 EMTs, and 30 firewomen. Around 1983, a third second-story addition was built that included a dorm, a kitchen, a meeting room, and a bathroom. (Courtesy of Barnardsville Fire Department.)

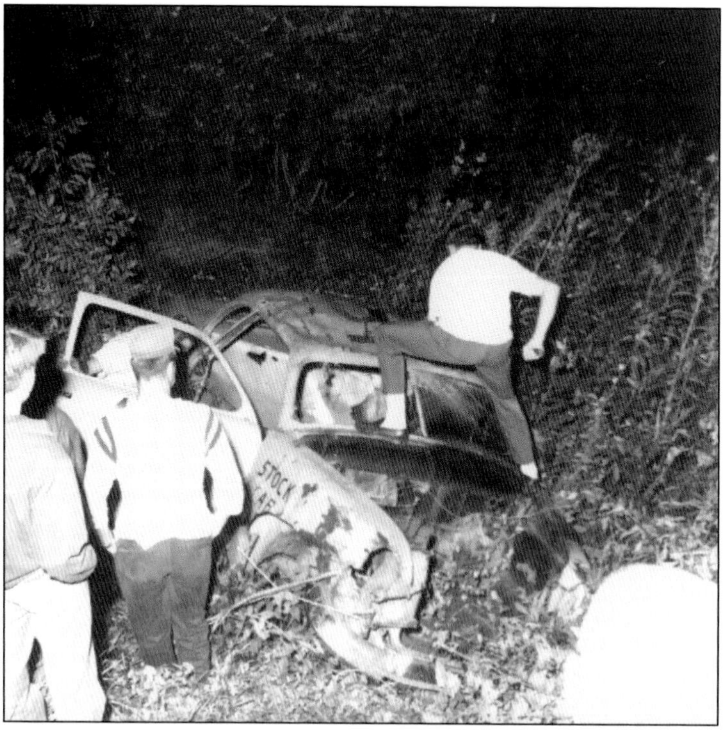

In September 1964, the rescue squad completed training from the North Carolina Department of Insurance, which included the methods seen above for removing victims from crashed automobiles. In 1990, the first paid members were hired. The Barnardsville Fire Department's new station answered its first call on October 9, 2006. Roger Maney started as a junior member in 1958 and is still serving today after over 40 years. (Courtesy of Barnardsville Fire Department.)

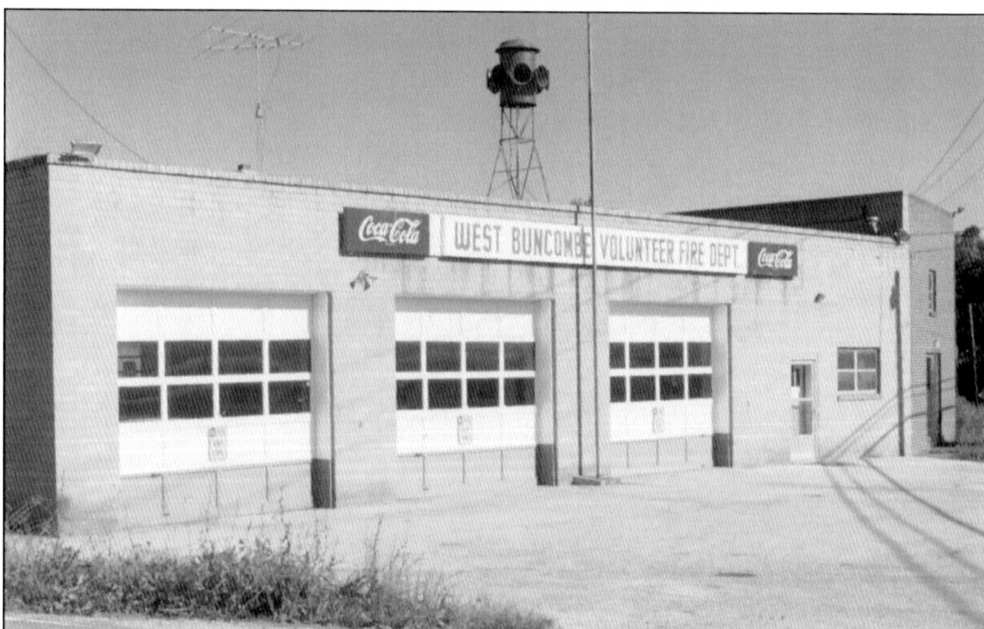

A group of citizens met at the Clyde A. Erwin High School on February 14, 1959, to form the Erwin Fire Department. G.R. Treadway was elected as the first chief, and 21 charter members volunteered. Before being chartered on March 10, the name was changed to West Buncombe Volunteer Fire Department because there was already a fire department in eastern North Carolina with the same name. A truck was ordered without a penny to pay for it. (Courtesy of West Buncombe Fire Department.)

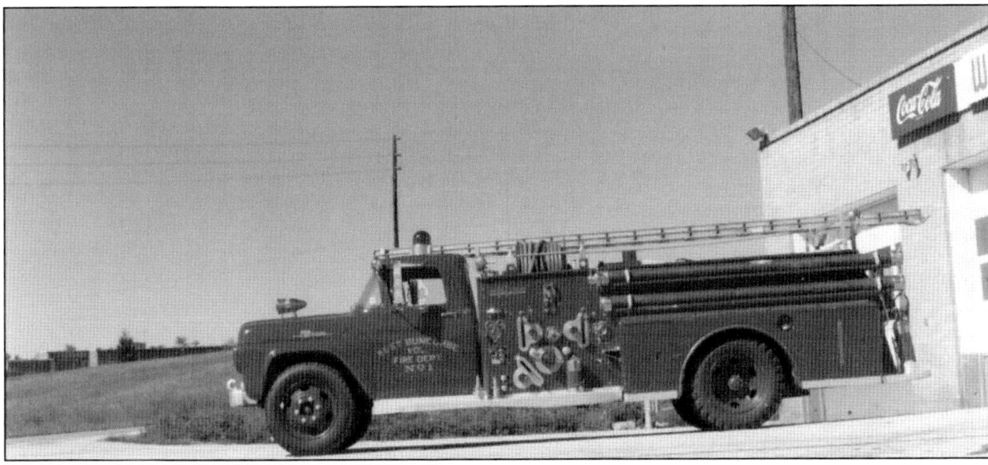

The money was raised by the time this $8,000 Ford American LaFrance 500-gallon truck was placed in operation in September 1959. Construction was begun on the two-bay station by February 1960 and was almost complete in December. Residents of the fire district were encouraged to donate $5 apiece. A few years later, one bay and an office were added on. (Courtesy of West Buncombe Fire Department.)

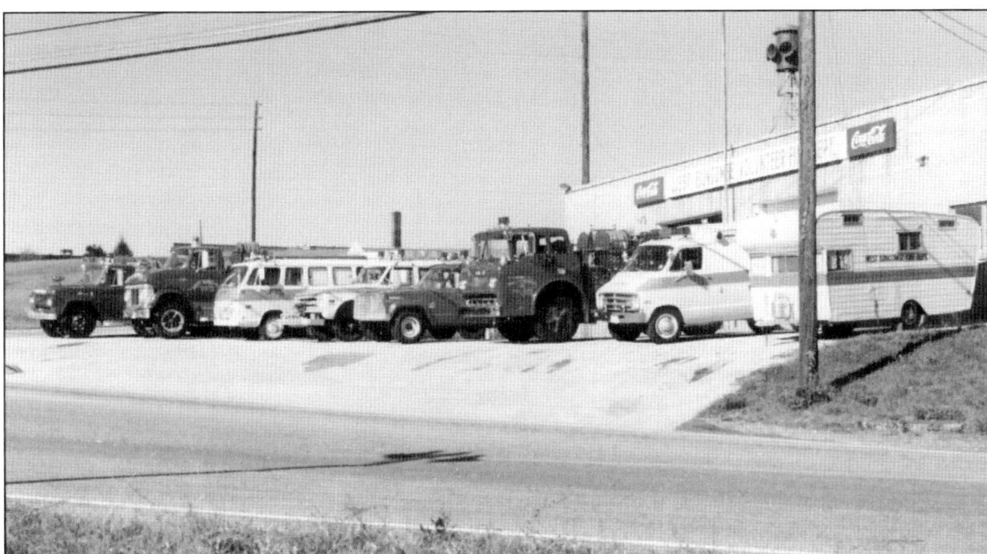

A fire one frigid night in February 1959 at Leonard Clement's house was fought by AFD and ENKA Fire Department. The next morning, 50 men showed up to start rebuilding. This blaze sparked the development of the West Buncombe Fire Department. A 1965 Ford, seen second from left, was the next truck purchased. The camper was used for a command post created in July 1976. (Courtesy West Buncombe Fire Department.)

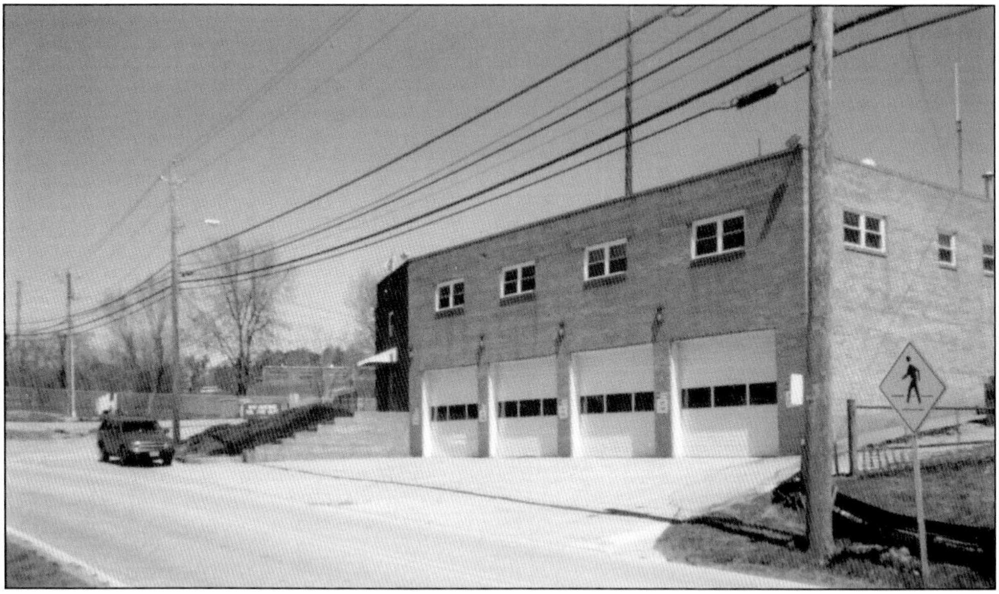

A third two-story, four-bay addition built in 1967 included a meeting and radio room. The office in the second addition was converted to a bay before 1979. A four-bay metal building was erected behind the station in the early 1980s, and a multimillion-dollar new station built directly behind the original opened on December 20, 2001. The original station was torn down by February 2002. (Courtesy of West Buncombe Fire Department.)

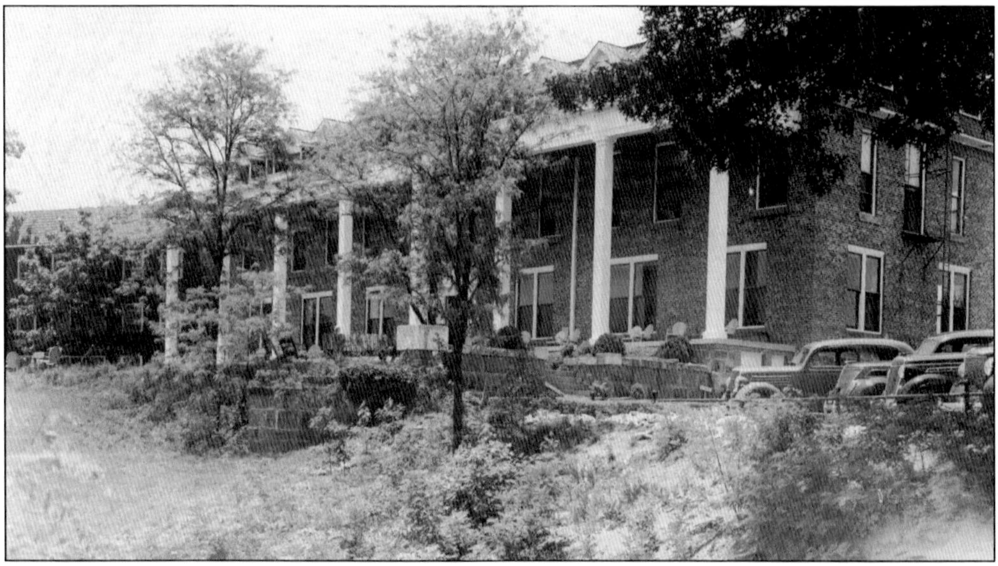

Ben Lippen School was started as a 65-acre interdenominational conference center in 1929 by the Columbia Bible College in Columbia, South Carolina. The private boy's boarding school was established at the site in 1940, becoming coeducational in 1950, and grew to a student body of 175 students from 26 nations and 23 states in grades 9 through 12. Most students were children of missionaries. The Inn, as it was known (pictured above), was built by the summer of 1933 using the bricks of the old 1892 federal post office and courthouse building that used to occupy the space now known as Pritchard Park in downtown Asheville. (Courtesy of North Carolina Collection, Pack Memorial Public Library, Asheville.)

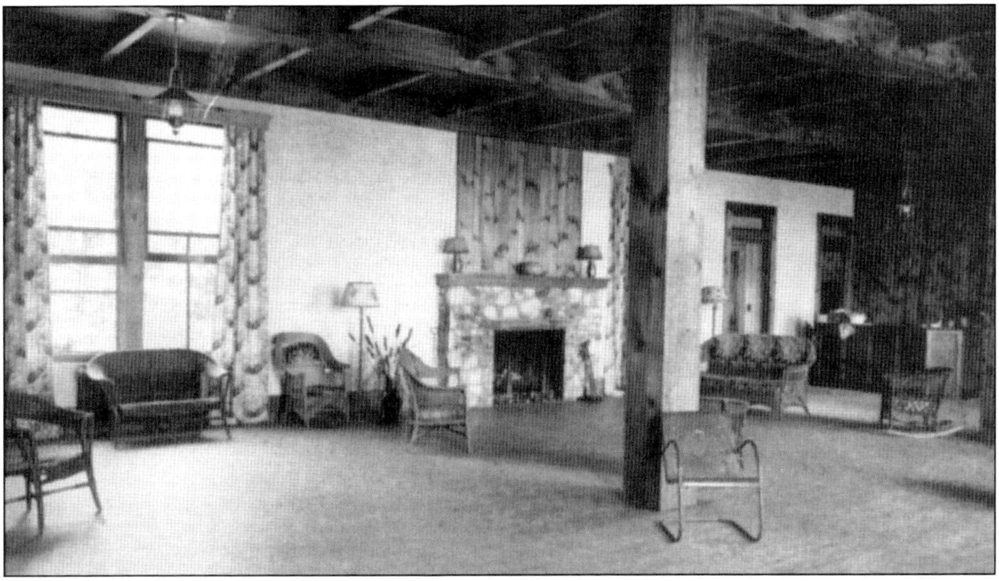

The building housed the student store and workshop in the basement, the kitchen and dining hall on the first floor, the storage and music rooms on the second floor, and dorm rooms for 18 mostly freshmen girls on the third floor. On the night of December 11, 1980, a fire was made in the fireplace after a recent $425,000 renovation and just five days before Christmas break. (Courtesy of North Carolina Collection, Pack Memorial Public Library, Asheville.)

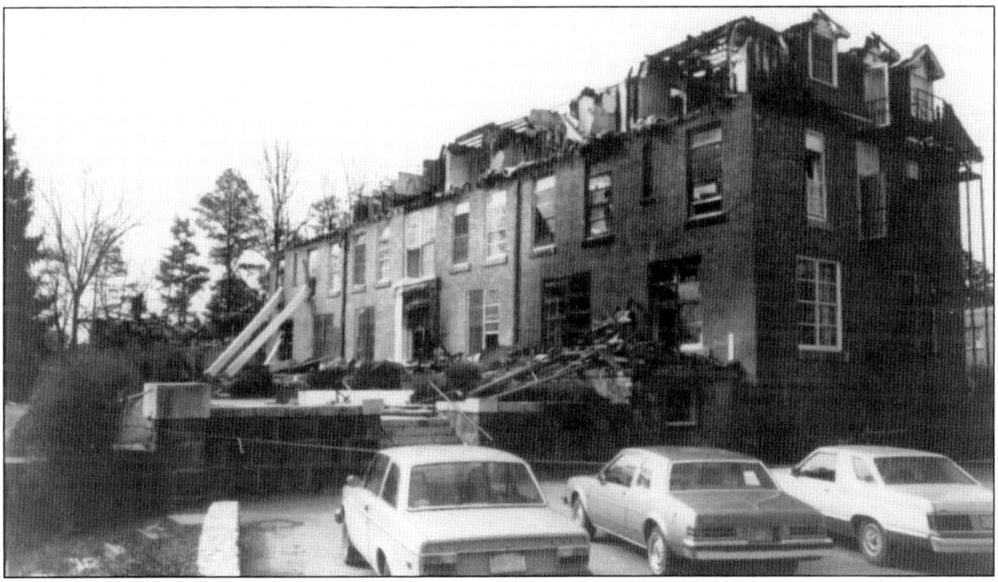

The next day, around noon, the fire alarm went off just before lunch. West Buncombe Fire Department responded to find smoke coming from third-floor cracks and vents and the area around the main chimney. For 30 to 45 minutes, firefighters searched for the blaze that was growing between the ceiling and the roof. Once an inspection opening was made, the fire flashed and broke through the roof around the chimney. The new sprinkler system, which did not cover the attic space, could not keep up, and the flames burned out of control for over four hours. The school moved to the Columbia campus in 1988. (Courtesy of North Carolina Collection, Pack Memorial Public Library, Asheville.)

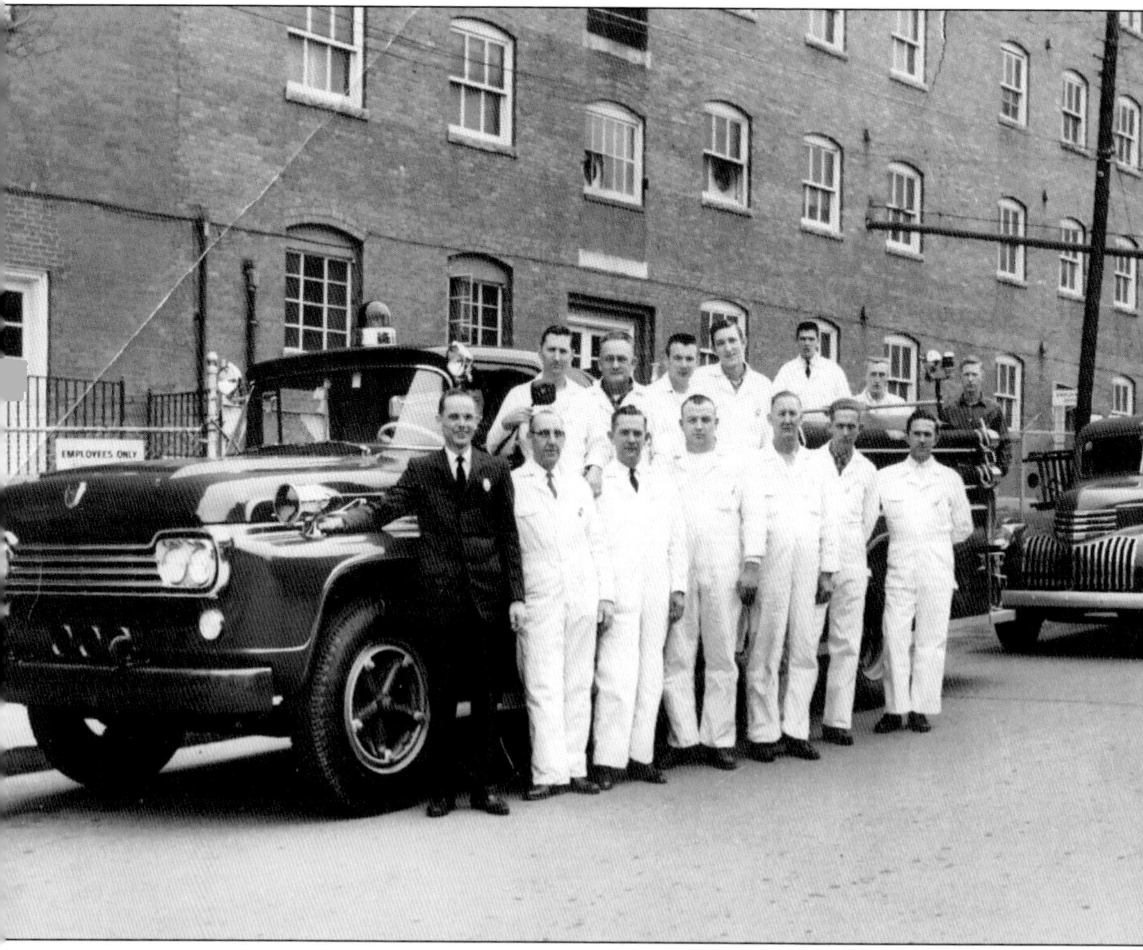

Swannanoa Volunteer Fire Department started when the area grew to be too much for the Beacon Fire Brigade to handle. (The Beacon Manufacturing plant is in the background). The department was chartered on April 2, 1959. Pictured above, the new $11,996.88 Ford American LaFrance with a 500-gallon tank arrived in July and was funded by the community through the Swannanoa Men's Club and Beacon. The truck to the right is a Chevrolet. Horace Stroupe was the first chief. (Courtesy of Swannanoa Fire Department.)

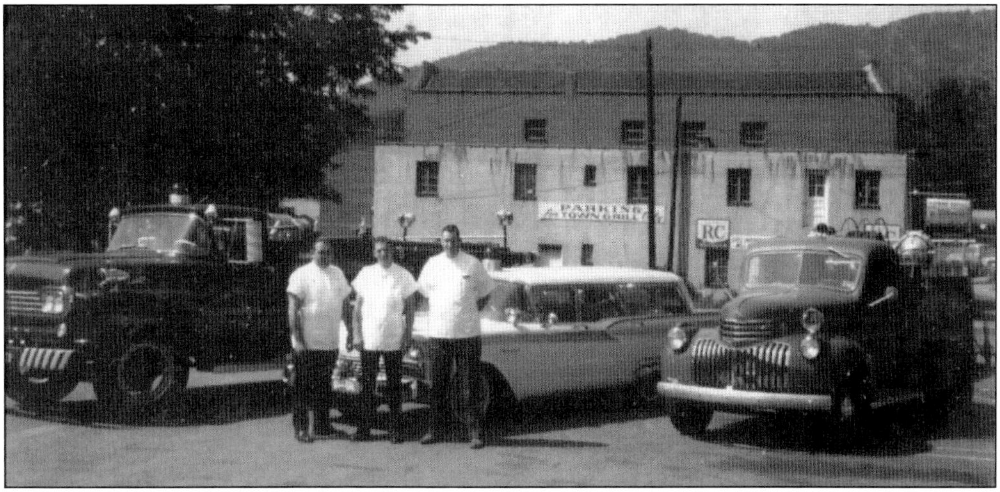

The fire department moved into the first two-bay station in 1961. The site was an old diner in the J.T. Mayfield building across from the plant. One volunteer was on duty every night at the station. In September 1963, a fire started in the recreation hall and faculty quarters at the North Carolina Juvenile Evaluation Center. The sprinkler system was turned off after a pipe broke during a freeze the previous winter. Another fire occurred in February 1976 at 2:12 a.m. The fire was through the roof of the vocational shop building upon the responders' arrival, and they fought it for over six hours. Black Mountain responded, but they had been hard at work burning a house for training until 11:00 p.m. earlier that night. These buildings were part of the old Moore General Hospital. The property was transferred to the North Carolina Board of Correction and Training, and the Juvenile Center opened July 1, 1961. (Above, courtesy of Swannanoa Fire Department; below, courtesy Charles George VA Medical Center.)

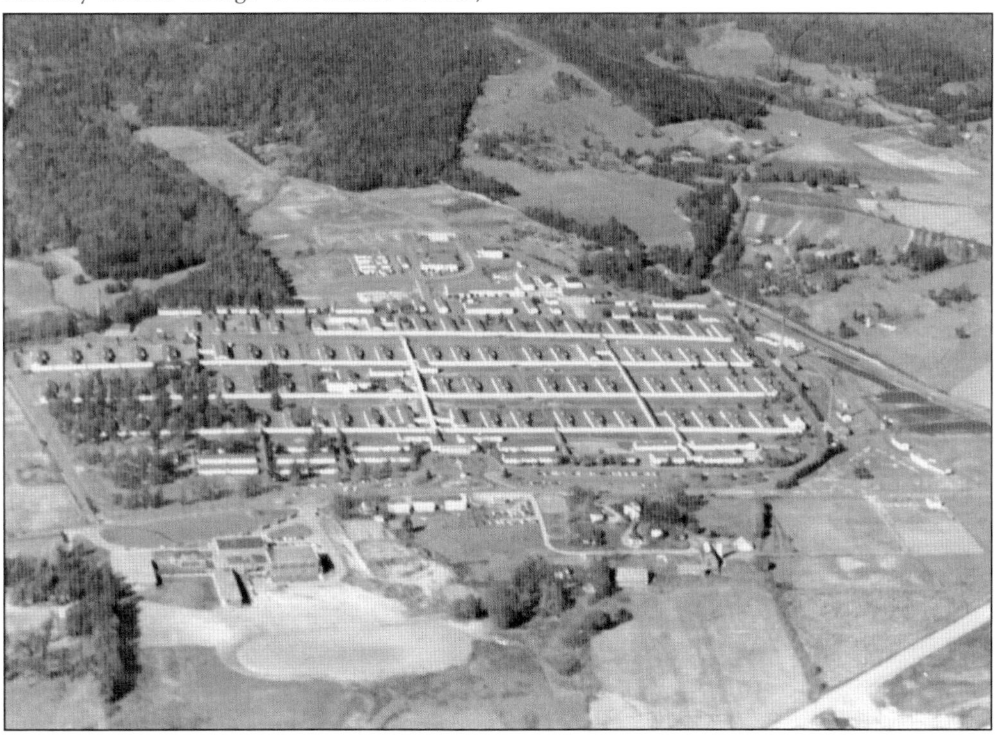

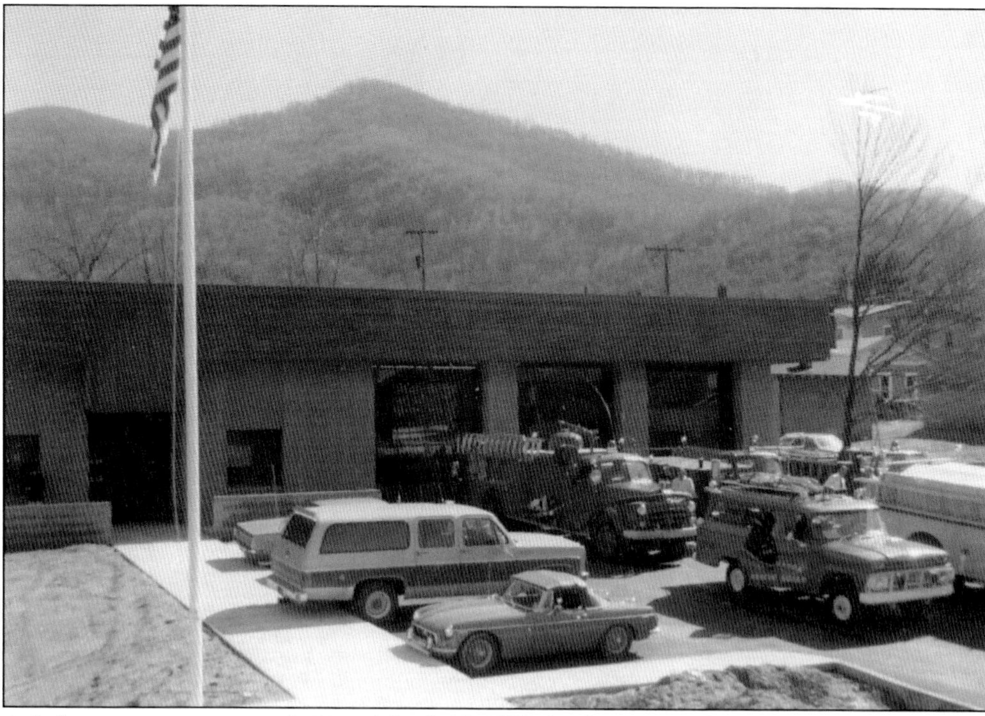

A dedication ceremony was held on March 13, 1977, for this new six-bay station built next to the first station. The facility was paid for through a $330,000 Farmers Home Administration loan. Equipment and two trucks were also bought; a Ford American LaFrance 1,200-gallon kerosene-powered pumper-tanker arrived in May 1976, and an American LaFrance Pioneer pumper arrived in August. Beacon's former foam truck can be seen in the center of this photograph. (Courtesy of Swannanoa Fire Department.)

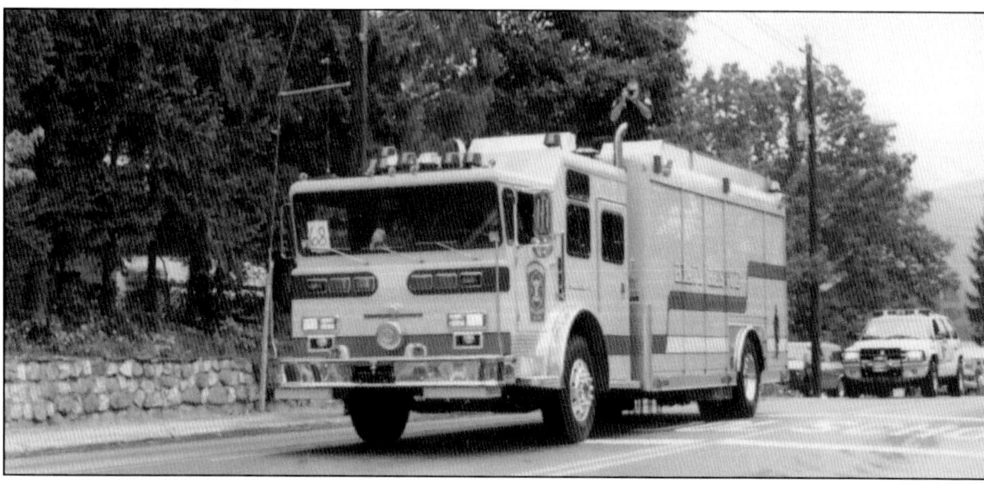

In December 1977, Barbara Suttle was one of the first female firefighters in the area. Swannanoa also had two female rescue squad members. Riceville had a female ambulance driver, and Skyland was training a female through Asheville-Buncombe Technical Community College (AB-Tech). Upper Hominy had two female ambulance attendants, Ruby Green and Argie Warren, in December 1972. The 1976 Pioneer was converted into a rescue truck in 1998 (above). (Courtesy of Swannanoa Fire Department.)

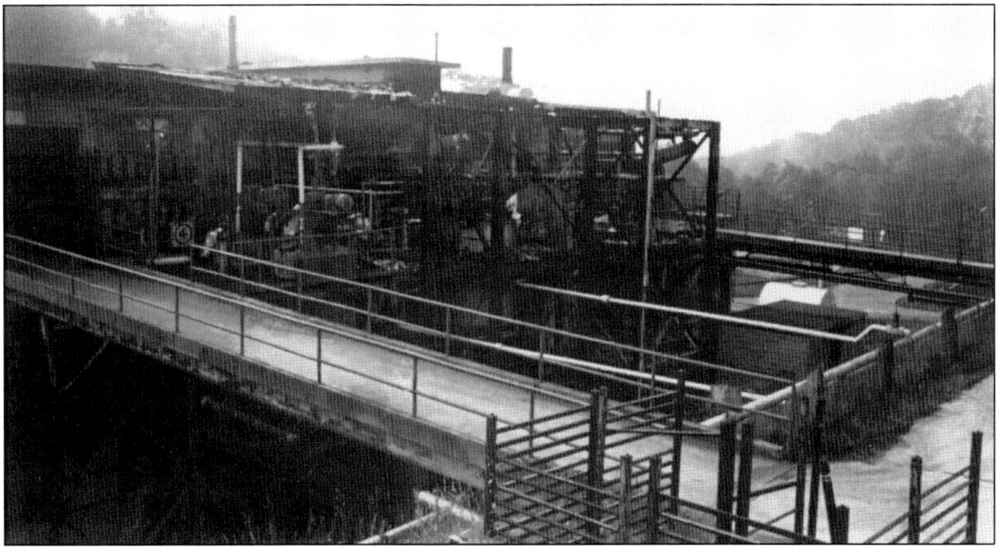

On July 7, 1965, a fire at one of 27 buildings on Amcel Propulsion Company's 1,300-acre plant threatened 3,000 pounds of explosives, causing around 2,000 people in a two-mile radius to evacuate. On May 12, 1986, at the now Chemtronics Inc., a reactor burped hot acids, burning the walls of a building, as seen above. Chemtronics was not cooperative in sharing information, and after this call, a new law started on May 25, 1986, requiring companies to share information on hazardous materials. In March 1990, a building at the now Jet Research Center containing up to 1,700 pounds of explosives caught fire. This time, the company was very cooperative. (Courtesy of Swannanoa Fire Department.)

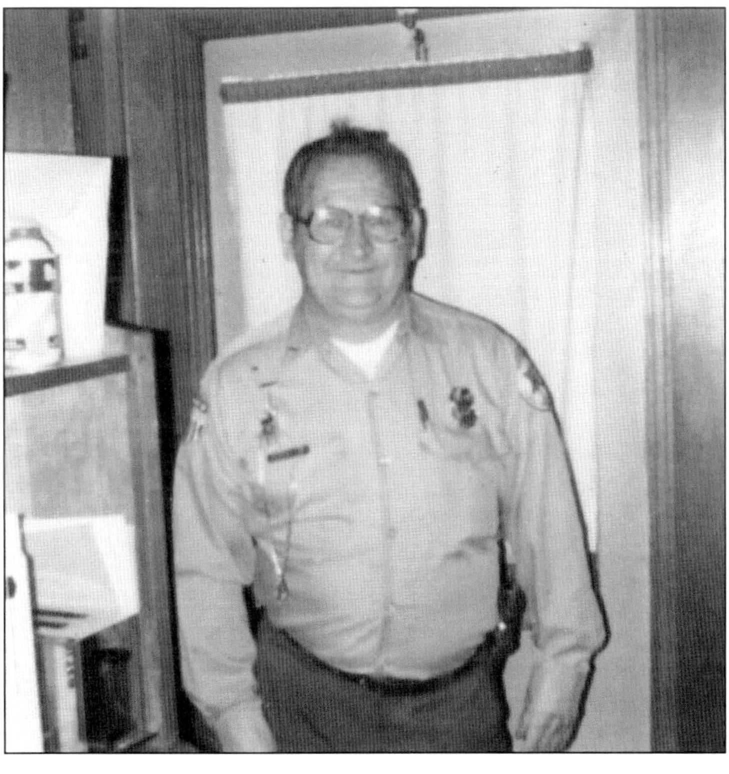

Sammie J. Waddell responded to a call during a snowstorm for subjects trapped on Interstate 40 near the 57-mile marker just east of the Interstate 240 interchange. A tractor-trailer jackknifed on the ice-covered road and slid across the pavement, striking the first responders on scene and their vehicles. Sammie was taken to Mission Hospital, where he died at 8:10 a.m. on February 23, 1989. He was 56 years old. (Courtesy of Swannanoa Fire Department.)

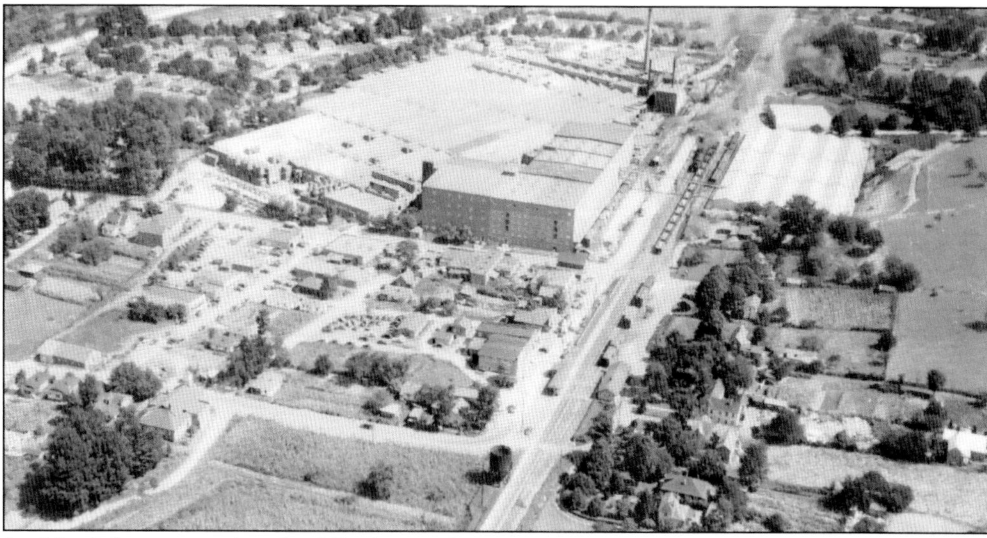

Paid firefighters started in the fall of 1988. On May 13, 1999, the board voted to build a separate three-bay addition to the station. On January 19, 2003, Warren Wilson College's Schafer dorm burned. On September 3, 2003, at 3:56 a.m., the largest fire in Buncombe County history occurred. A teenage arsonist had set fire to the nearly 80-year-old Beacon Manufacturing plant just 50 feet from the station. The mill had sat empty since its closing, and one million square feet (18 acres under one roof) of one-, four-, and six-story interconnected buildings had their utilities and sprinkler systems turned off. (Courtesy of North Carolina Collection, Pack Memorial Public Library, Asheville.)

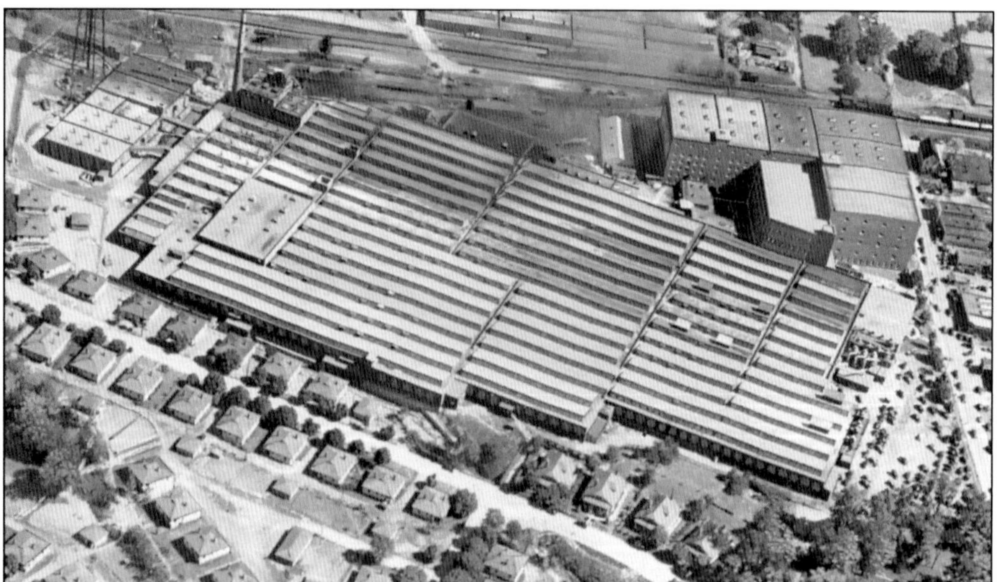

The fire department had decided after a preplan that if the fire were easy to reach, they would fight it, but if it were not (and this one was not), they would not risk going through the maze of hallways to reach it. Large holes three to four feet deep were left in the floors where machines used to sit. The plant buildings were mostly brick, some with heavy timbers, and they held old machines and leftover cloth. (Courtesy of North Carolina Collection, Pack Memorial Public Library, Asheville.)

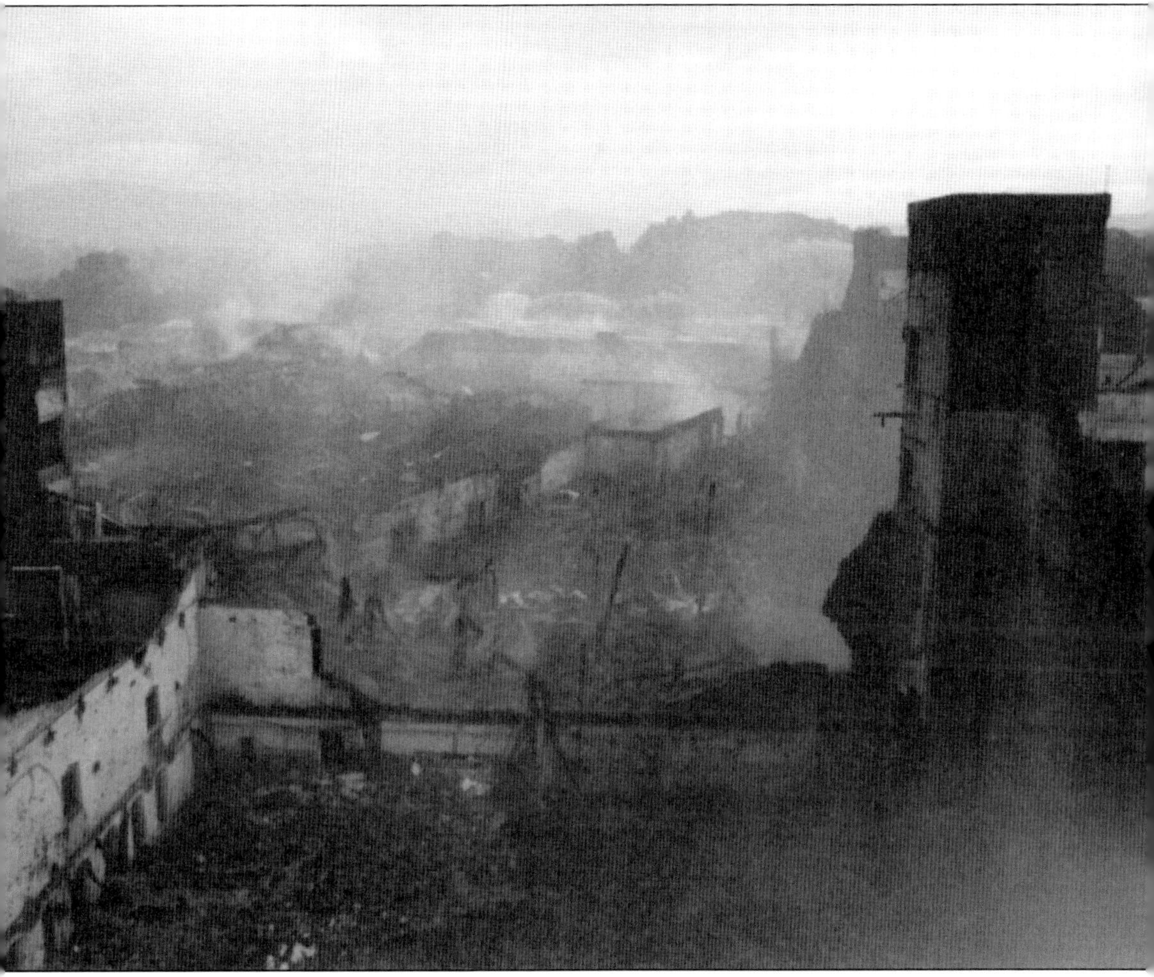

Within minutes of the firefighters' arrival, portions of a four-story building were collapsing. A total of 32 fire departments and over 500 people responded to help contain the blaze, which burned 24 buildings for 112 hours. Nearby fire hydrants only supplied 500–1,000 gallons per minute, so a trench was dug under the nearby railroad tracks to lay a four-inch line, water was relayed from the nearby Swannanoa River, and seven tankers were used to shuttle water all for the seven aerials and 24 pumpers, which used 25 million gallons of water in the first 12 hours. (Courtesy of Asheville-Buncombe Arson Task Force.)

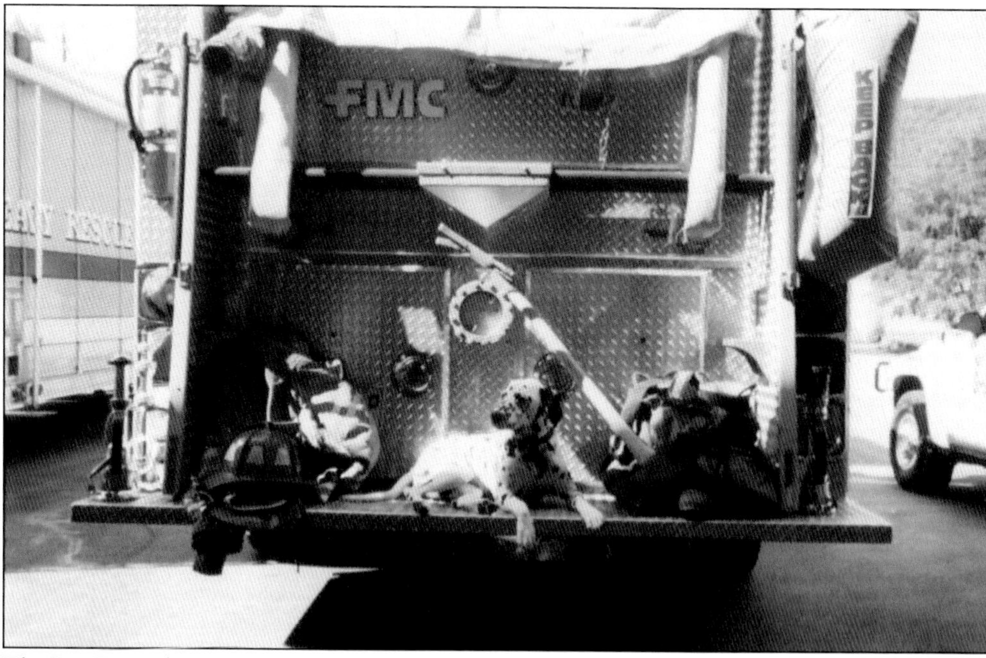

The Bee Tree substation opened in 2006. Dottie LaFrance was the first station Dalmatian donated to the fire department. Ashes was next, and Bell was the last. The dogs lived at the station and went to fire-prevention programs, and Ashes was even trained to stop, drop, and roll. Dottie and Ashes were given a fire-service funeral and are buried with toys in front of the station. (Courtesy of Swannanoa Fire Department.)

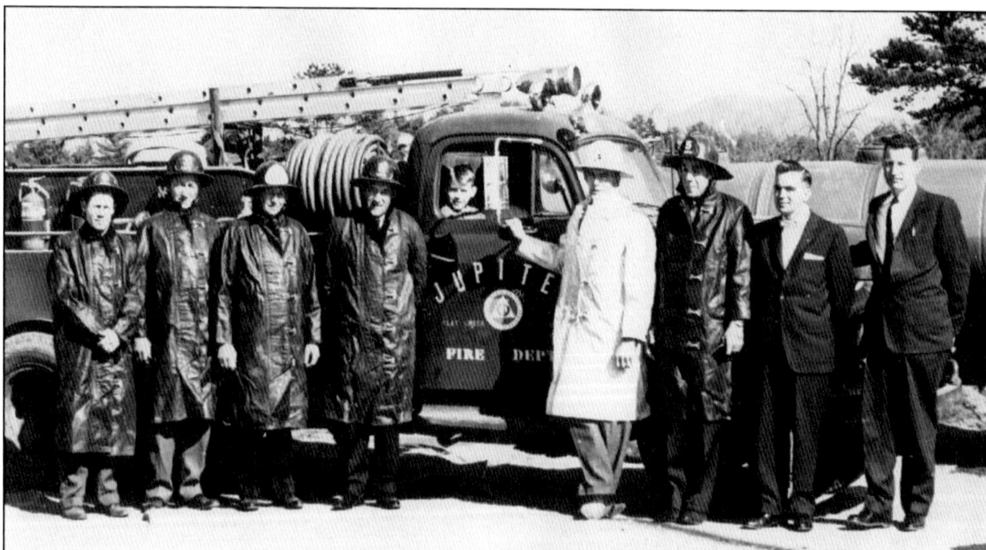

Jupiter Volunteer Fire Department was founded as the Red Oak, Jupiter, and Flat Creek Fire Department on August 27, 1959, at a meeting of the Red Oak Wildlife Club. E.B. Debruhl was the first chief. The fire department was started to protect the large chicken industry in the area. The chicken litter and dust created a potentially explosive combination with the wood stoves and gas heaters used to heat the chicken houses. (Photograph by June Glenn Jr.; courtesy of Jupiter Fire Department.)

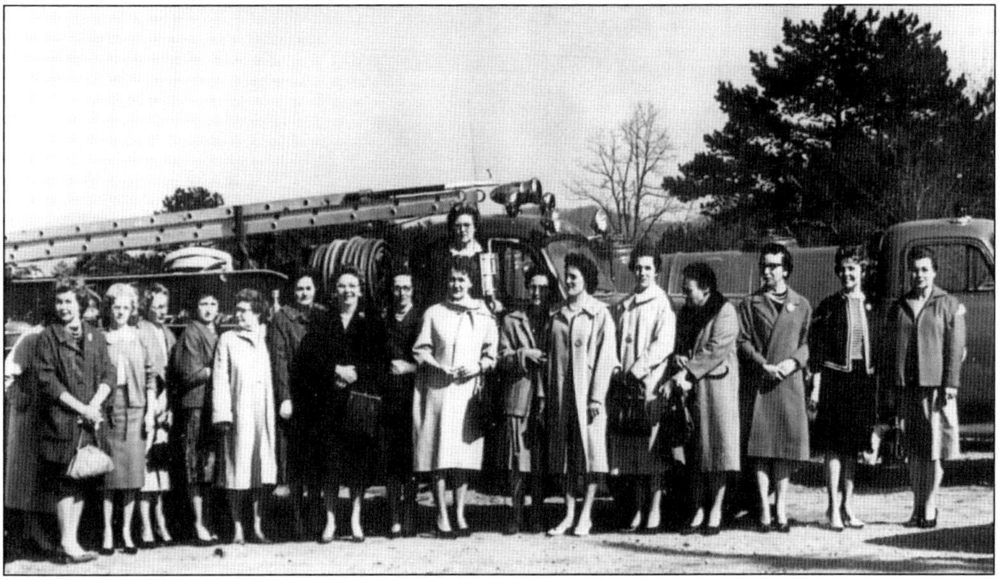

The first truck was built from a 1951 Ford Army surplus vehicle from Moore General Hospital using metal signs. It had a 300-gallon-per-minute pump, a 780-gallon tank, and a new type of extinguisher that used carbon dioxide. It was completed December 19, 1959. A 1952 Chevy 1,080-gallon tanker truck was built from an old surplus truck and an oil tank bought from West Buncombe Fire Department in 1961. (Courtesy of Jupiter Fire Department.)

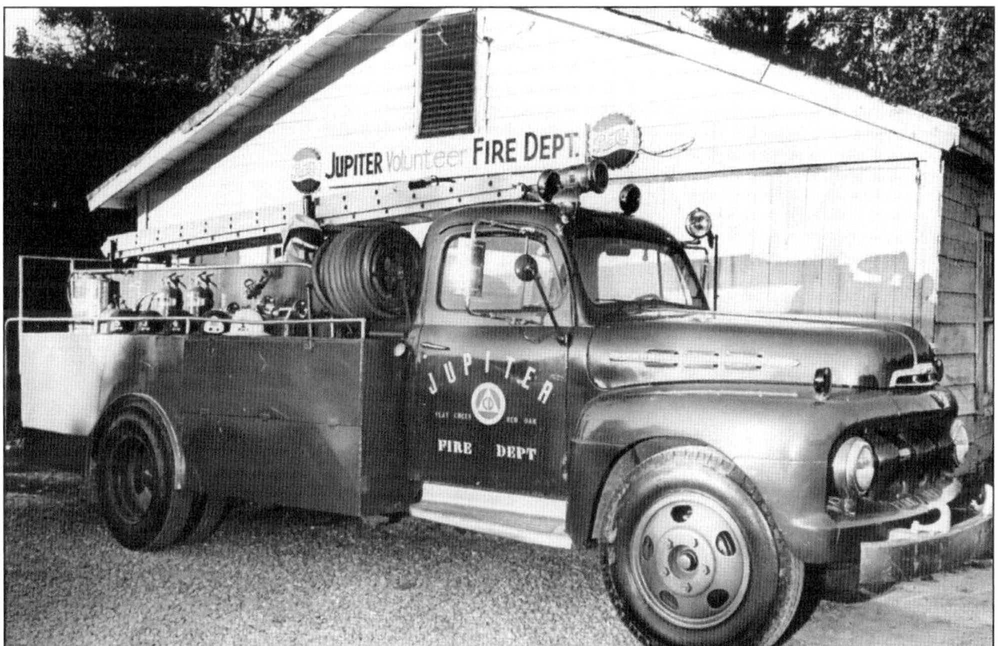

The truck was built at Holcombe's Service Station before being moved into a garage owned by Frank Reese in December 1959. Tarpaper and an oil stove were added to keep the truck from freezing. The next firehouse was in the old Crowder Printing Press (shown above), purchased July 12, 1960. (Courtesy of Jupiter Fire Department.)

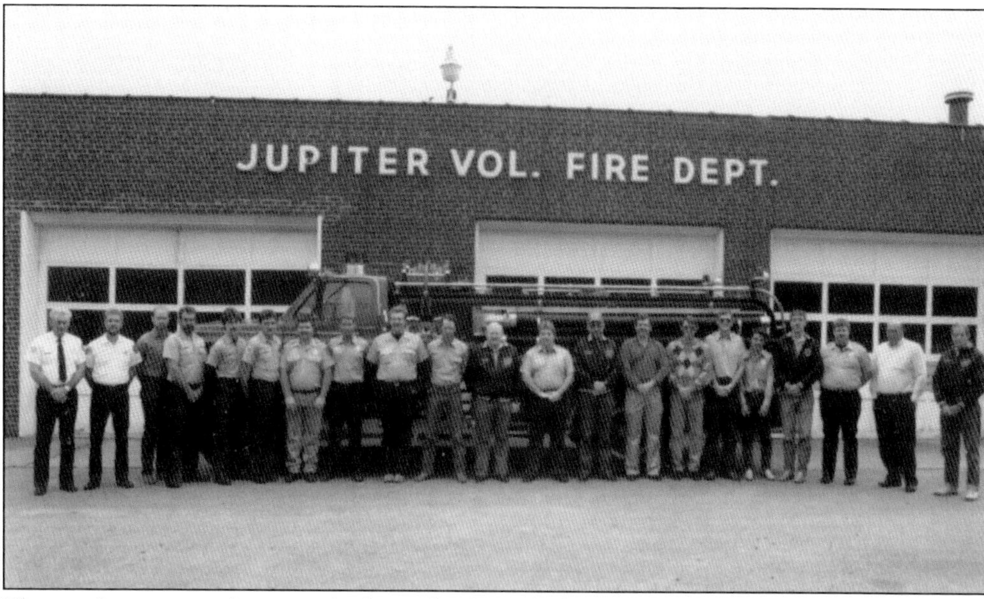

The two-bay station above was built on the Crowder lot in 1966. In the late 1960s, the broiler chicken industry ended in the area. They were contract growers for Earle Chesterfield Mills in Asheville, which burned in a spectacular fire on April 2, 1995. In 1984, the meeting room was converted to a third bay. In 1992, Jupiter moved into their new station. Their first paid firefighters were hired in 1999. (Courtesy of Shannon Roberts.)

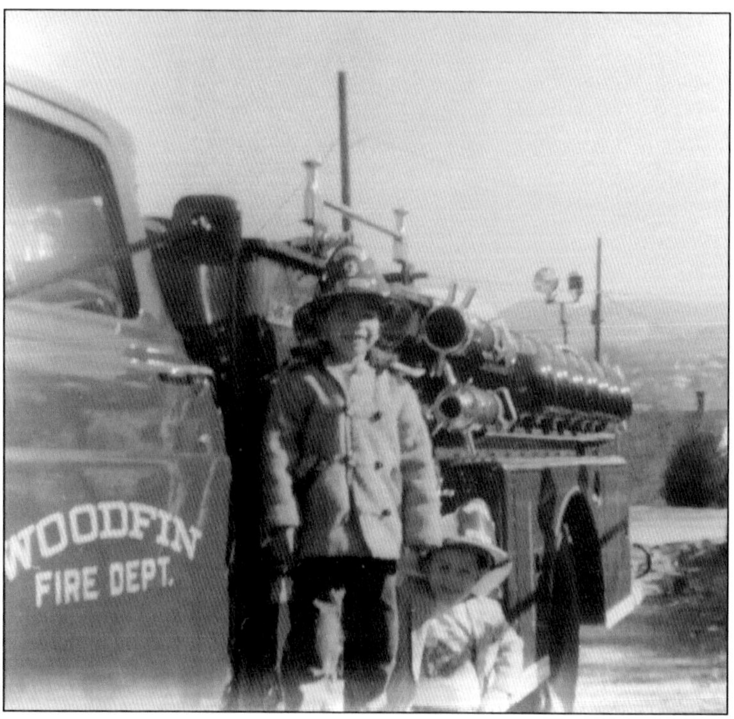

A community meeting was held on October 23, 1959, in the Woodfin School auditorium to organize the Woodfin Fire Department. Ernest Ellis was elected chief. The idea of a new fire department was conceived by the Woodfin Lions Club, and the meeting was held because AFD would no longer be responding outside the city limits after January 1, 1960. The first truck, an $11,500 American LaFrance, arrived February 1960. (Courtesy of Woodfin Fire Department.)

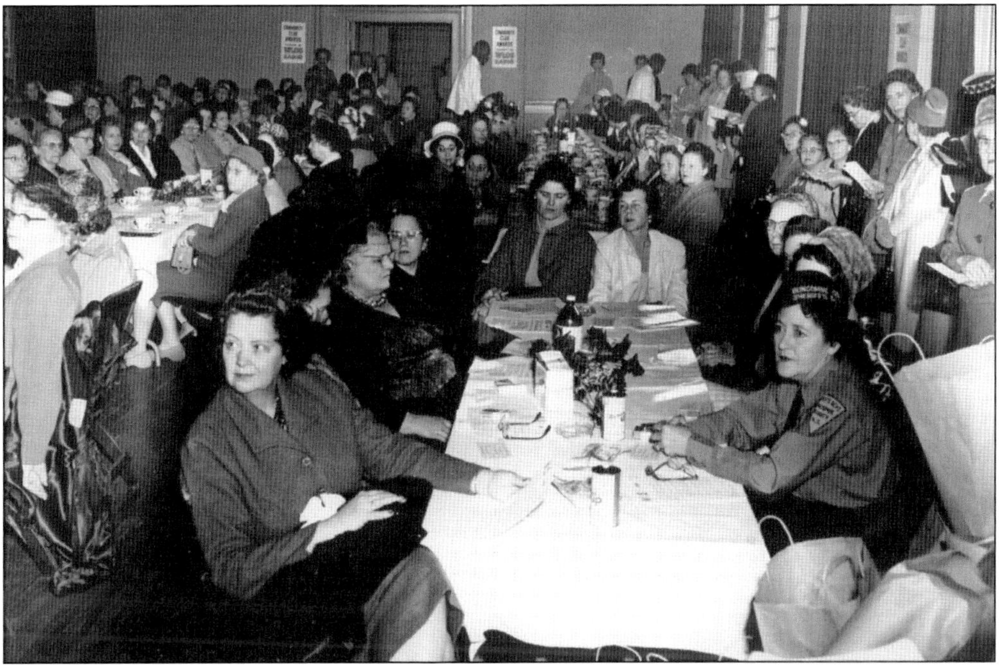

Members of Ladies Auxiliaries attended the WLOS Community Club Awards ceremony in May 1962. Money was raised during Christmastime at a country store located next to the Elkwood station, through the sale of drinks during elections, and with a scrap drive in which a 14-year-old boy won a bike for his efforts. They even sold a heater that had been brought by a fireman to heat the station. The group also produced the *Fire Fighter* newsletter. (Courtesy of Woodfin Fire Department.)

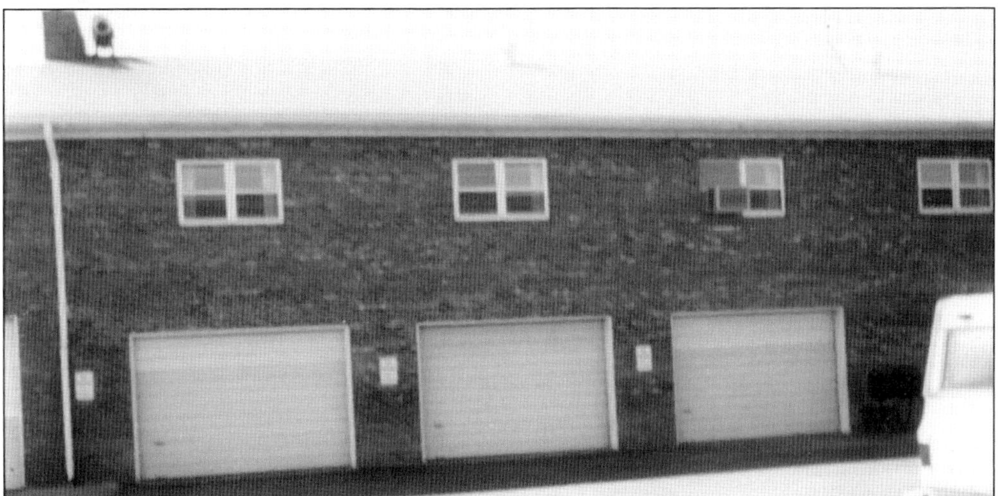

The first station was a rented building at 128 Elkwood Avenue. In October 1966, seven acres of land were given by Burlington Industries (a major textile mill that included a mill village and closed in 1985) for a community center and fire station. In 1968, the three-bay fire station (above) opened, complete with meeting room and offices. Upstairs was the community center with kitchen. (Courtesy of Woodfin Fire Department.)

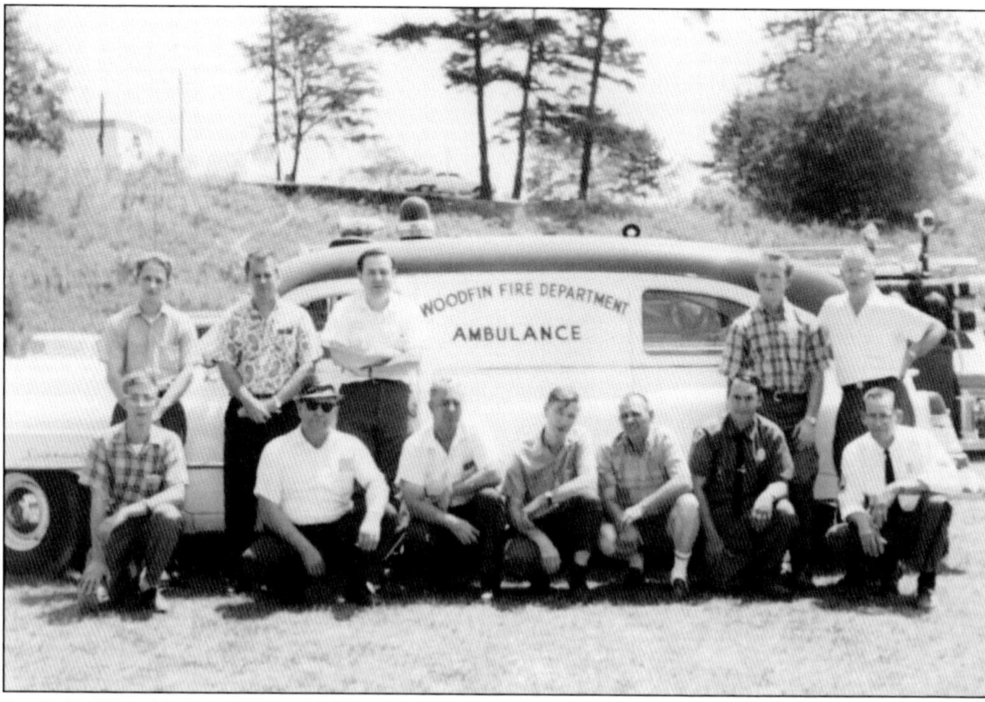

In 1969, Woodfin announced they would offer free ambulance service and bought the above ambulance. Also in March 1969, the community voted for a fire tax. To prevent annexation by the City of Asheville, the community voted to incorporate in January 1971. In April 1972, a dedication was held for the completed community center, town hall, and two-bay addition to the fire station. (Courtesy of Woodfin Fire Department.)

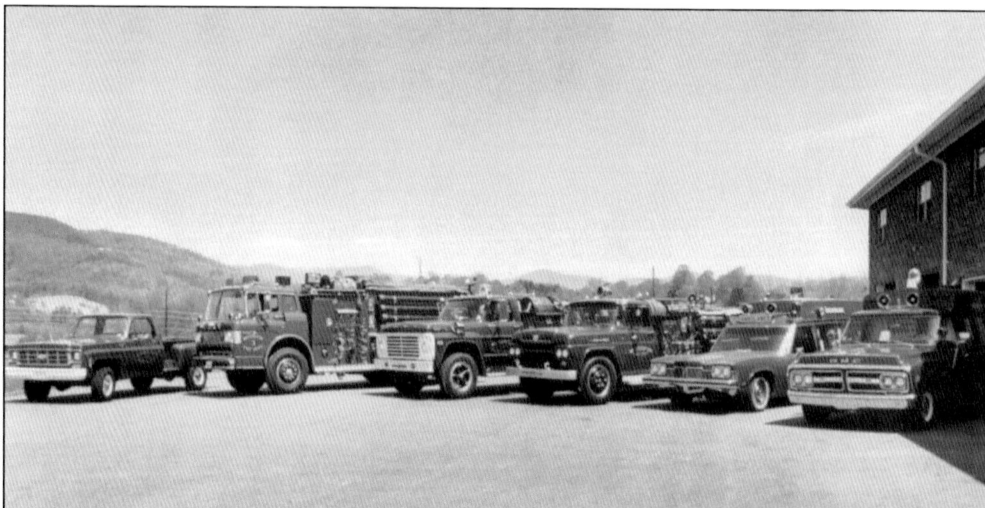

A $24,000 Ford fire truck (third from the left) was added in 1971. On January 16, 1973, a new $15,000 Pontiac Superior Coach ambulance was purchased. In 1975, the department had one full-time firefighter/ambulance attendant that was an EMT on each ambulance run with a driver (after July 1976, the driver had to also be an EMT). In the beginning of May 1979, a new, $75,000, 1,000-gallon Ford Howe truck (second from the left) arrived. (Courtesy of Woodfin Fire Department.)

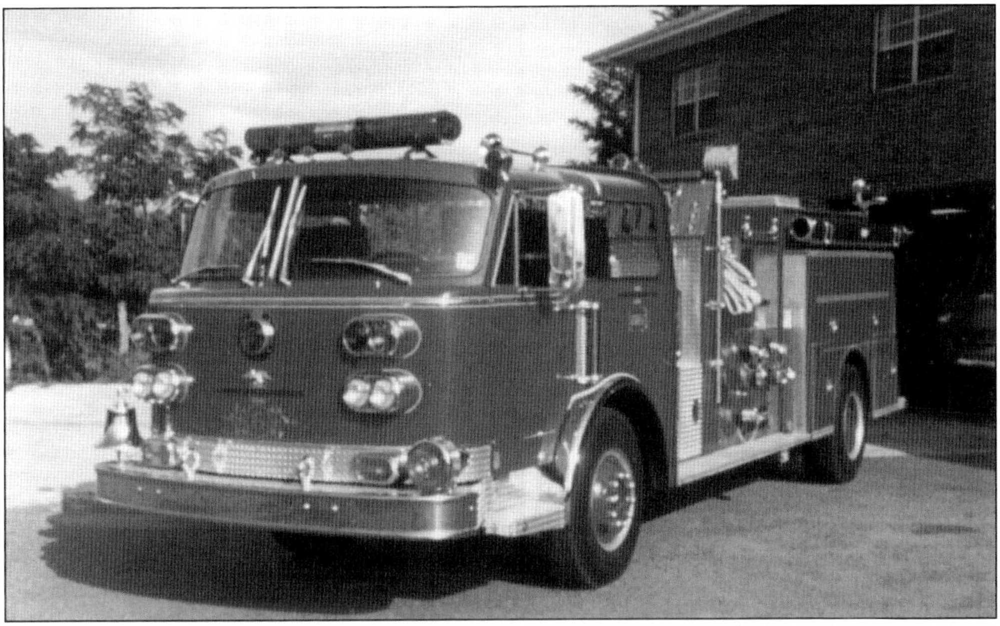

In September 1974, the area's costliest fire up to that time occurred at the Drexel furniture plant. The $9-million fire was responded to by 18 Buncombe County fire departments. In 1979, the Woodfin Fire Department had four sets of father-son volunteers, a common occurrence at fire departments. This new American LaFrance arrived in 1982. In 1996, the ambulance went out of service, and the department moved in to a new multimillion-dollar fire station. (Courtesy of Woodfin Fire Department.)

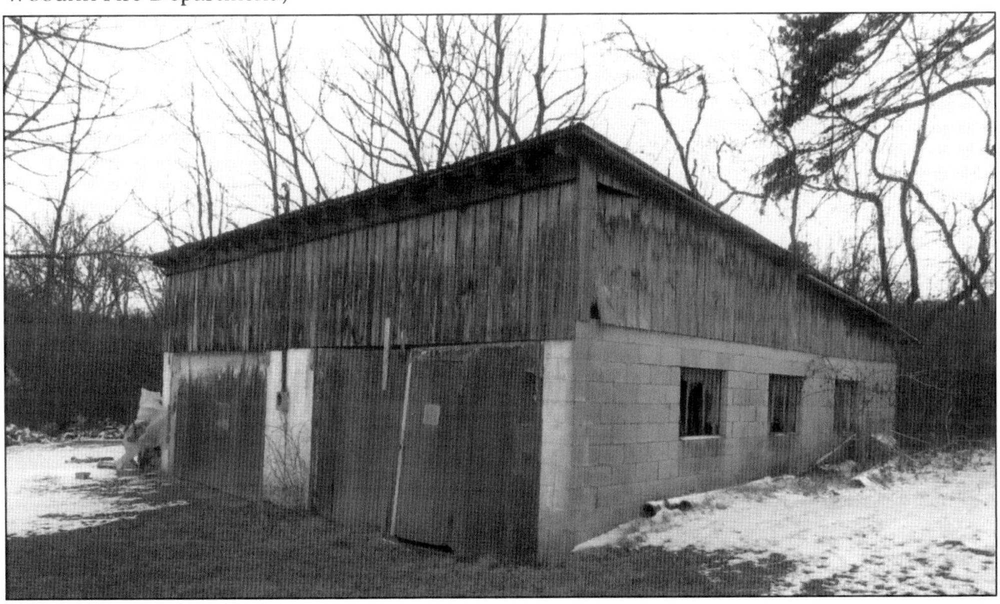

Upper Hominy Volunteer Fire Department was founded after a community meeting at Pisgah Elementary School on October 29, 1959. Ray Davis was named chief. The department was organized after Robert Foster Sr. lost his life in a house fire, and it was chartered on November 9. The first two-bay firehouse (above) was built behind Ray Davis's store and included a room for one fireman to sleep at all times. (Author's collection.)

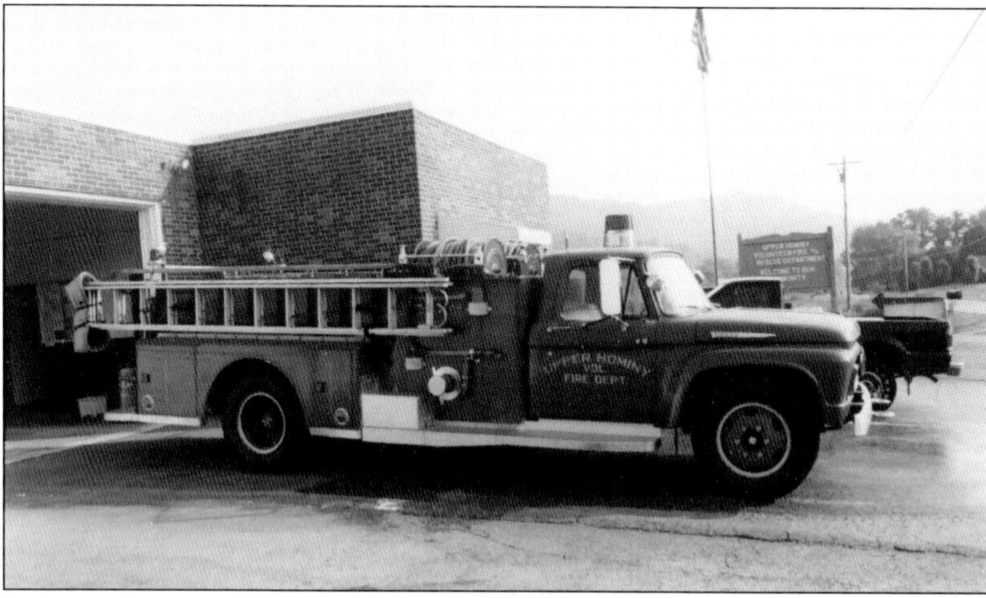

In January 1960, American ENKA sold the Upper Hominy department their first International 1,000-gallon truck. The first new $11,100 Ford American LaFrance 500-gallon-per-minute truck arrived February 1961. Every member of the community was asked to contribute $3 per building. In 1965, ENKA donated the first ambulance. In 1967, their second four-bay station was built, complete with meeting room, restrooms, and kitchen. A department-built, 1,200-gallon tanker went into service in March 1973. (Courtesy of Upper Hominy Fire Department.)

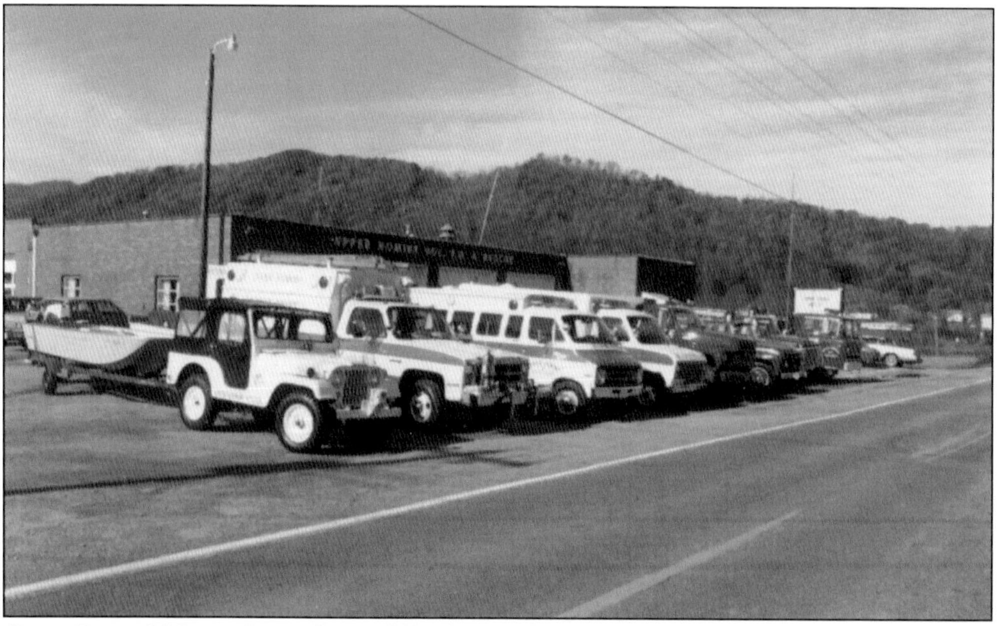

Groce Funeral Home donated a Cadillac hearse for an ambulance in May 1974. In 1975, an addition was made to the fire station that included a bay and office. In 1979, a new $53,000 Ford Grumman Howe was purchased and a tanker was built. The new Beaverdam substation had an open house on June 1, 1986. The department moved into a new multimillion-dollar station in 2010. (Courtesy of Upper Hominy Fire Department.)

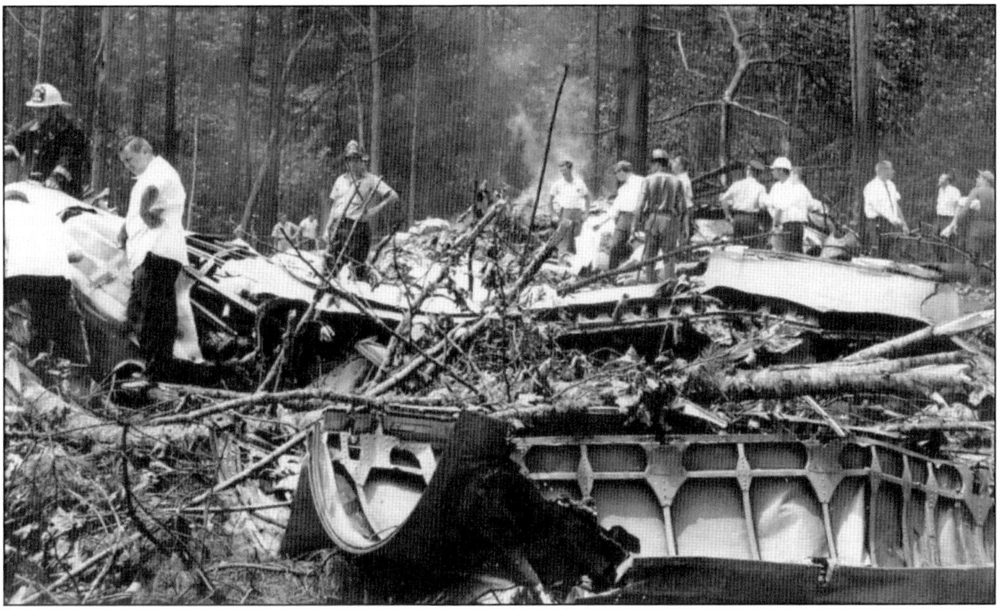

On July 19, 1967, a Piedmont Airlines Boeing 727 crashed midair above Hendersonville, North Carolina, into a Cessna 310, killing all 82 total occupants. It was the first plane crash investigated by the National Transportation Safety Board. Multiple rescue squads from Buncombe County, including Upper Hominy, assisted at the 1.5-mile-long, half-mile-wide crash site. It was determined that the Cessna did not follow its cleared approach flight path. (Courtesy of Upper Hominy Fire Department.)

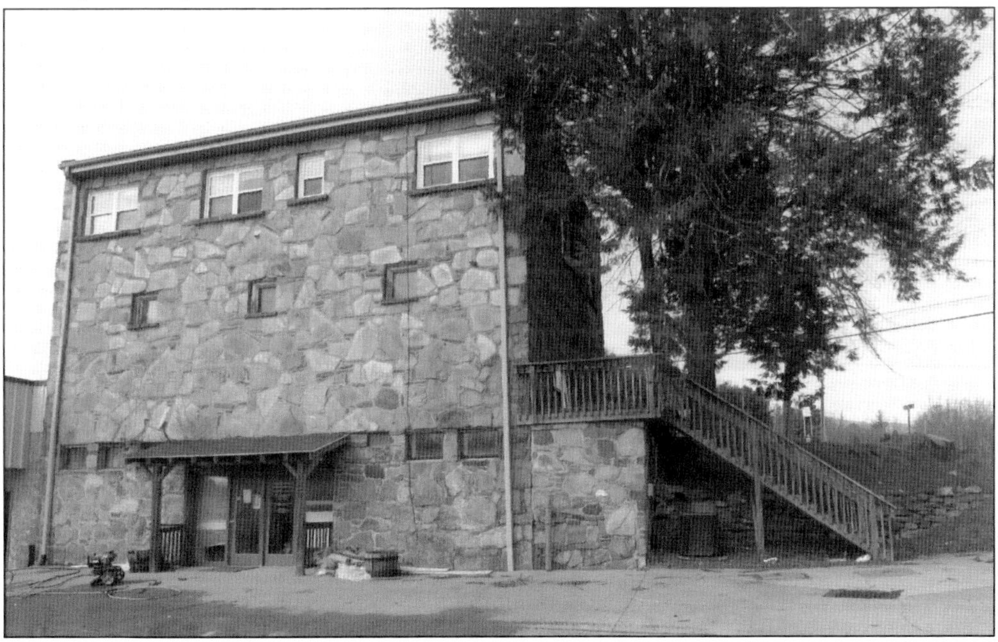

A group of seven community members met in the teacher's lounge at the Fairview School and gave $20 each to start the treasury for the Fairview Volunteer Fire Department on February 17, 1961. A.W. Nesbitt was the first chief. A temporary fire station was set up in the basement of the Fairview Feed and Grocery Store, shown above. (Author's collection.)

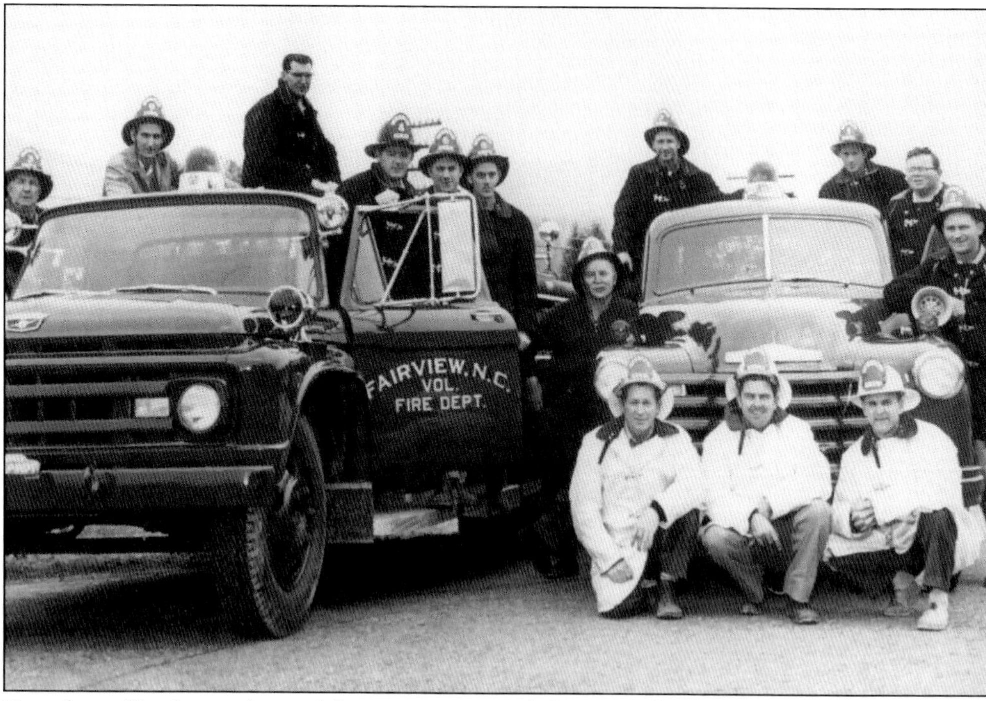

Two plots of land were donated for a station around the start of 1961. The 500-gallon American LaFrance fire truck on left, which cost $11,500, arrived in June, and on July 13, 1961, their department's charter was filed. The truck to the right was a 650-gallon tanker that the department built. The new fire station was dedicated on June 25, 1967. (Courtesy of Fairview Fire Department.)

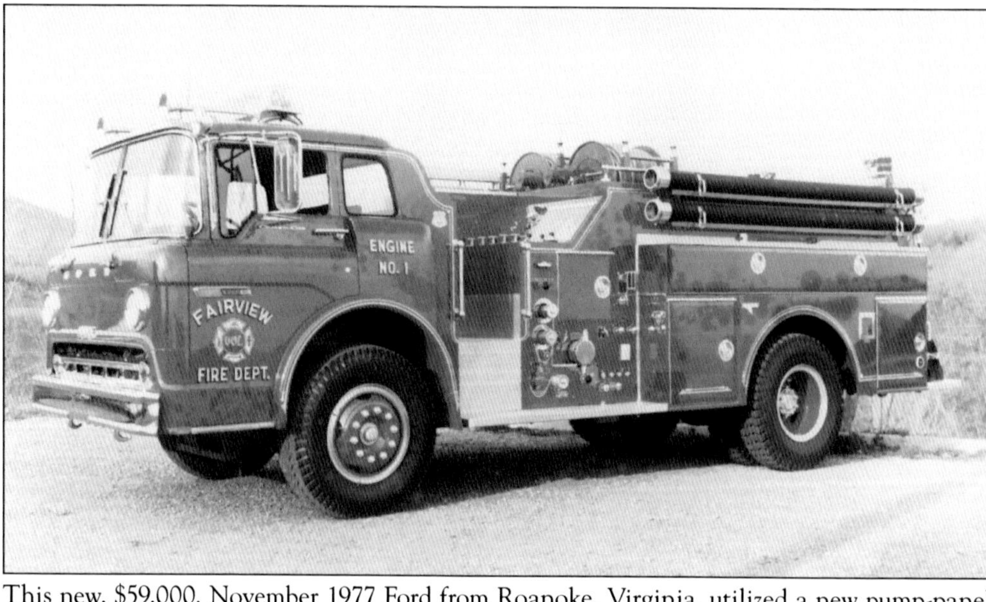

This new, $59,000, November 1977 Ford from Roanoke, Virginia, utilized a new pump-panel placement, allowing the engineer to see all around the fire scene. A fire tax started in 1963, but fundraising continued in 1978 with the raffling off of boats, trucks, steers, and hot dog and chicken dinners every other Sunday. A second three-bay addition was built around 1978 behind the original bays. (Courtesy of Fairview Fire Department.)

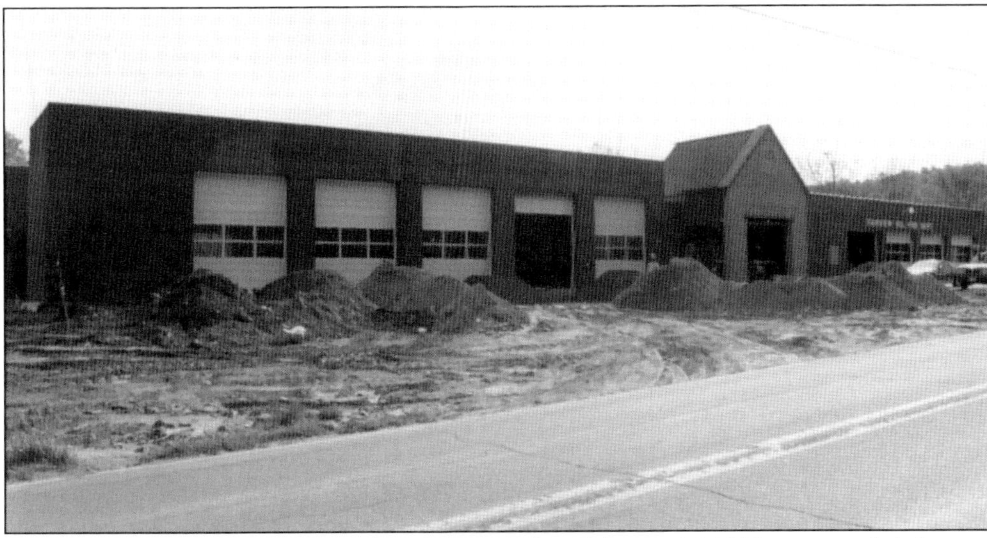

The board hired the first of two paid firefighters at their February 26, 1991 meeting. A dedication was held at the end of August 1997 for a third addition, shown above at left, complete with nine bays (bringing the total to 16), offices, a day room, a weight room, and room for sleeping quarters. In September 2007, the Old Fort Road substation opened. (Courtesy of Fairview Fire Department.)

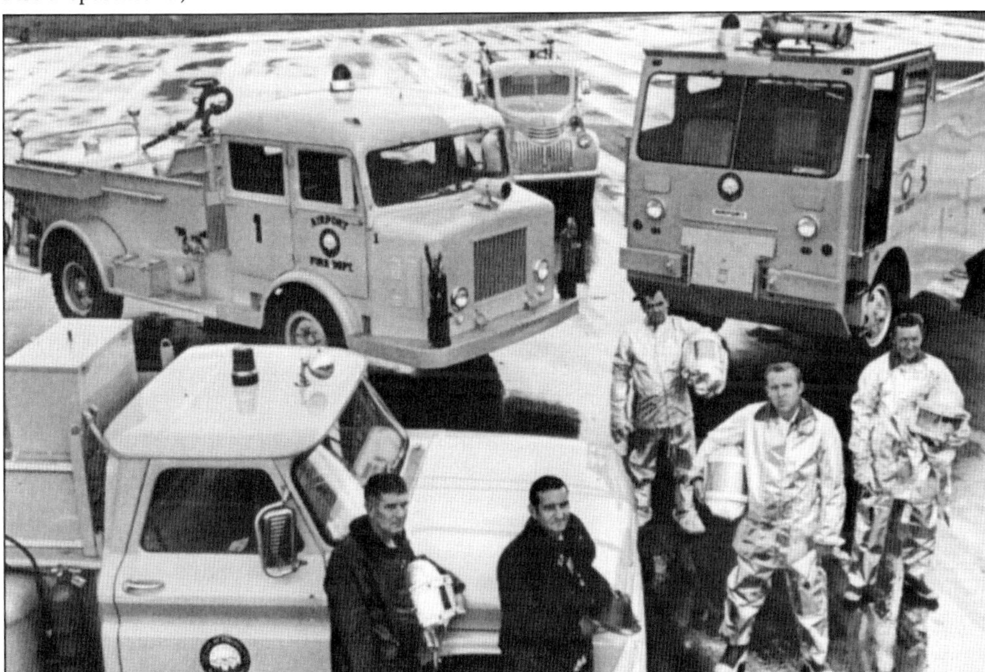

Asheville Regional Airport opened for service on June 7, 1961, replacing the Asheville-Hendersonville Airport that started in the late 1920s. The old airport's mountainous area could not support modern commercial planes. The Airport Fire Department started with one surplus pumper. Bill Lakenan was an early chief. In July 1967, the Piedmont Flight 22 crash showed a need for better equipment. In 1969, the department purchased two surplus water/foam trucks and modified the original truck to produce foam. (Courtesy of Asheville Regional Airport.)

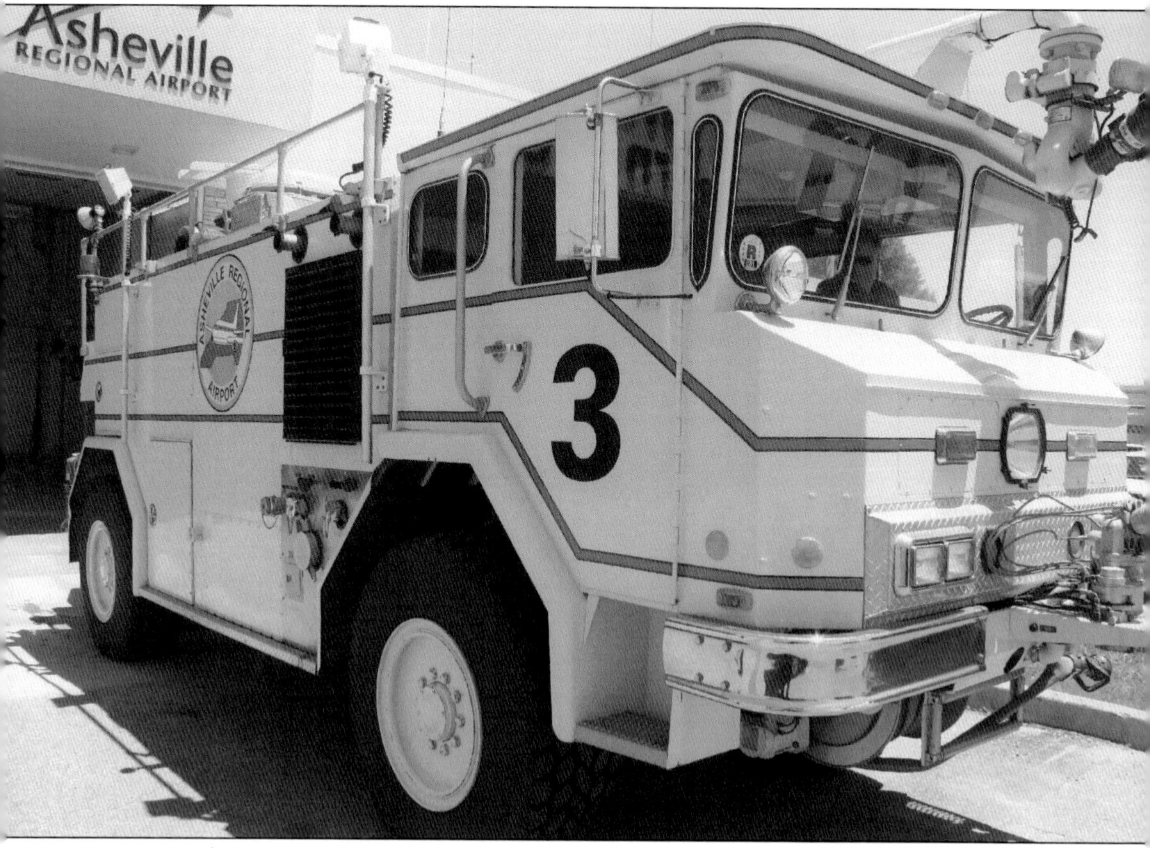

In 1975, the once separate Airport Police and Fire Departments were combined to form the Department of Public Safety (DPS), with members trained in both disciplines. Costing $150,000, the first new 1,500-gallon crash fire-rescue truck arrived in November 1977. This truck, lime-green in color, was a first for the airport; it had four-wheel drive, allowing it to go off-road, and it only took one firefighter to operate. It replaced two older trucks. (Courtesy of Jeff Augram.)

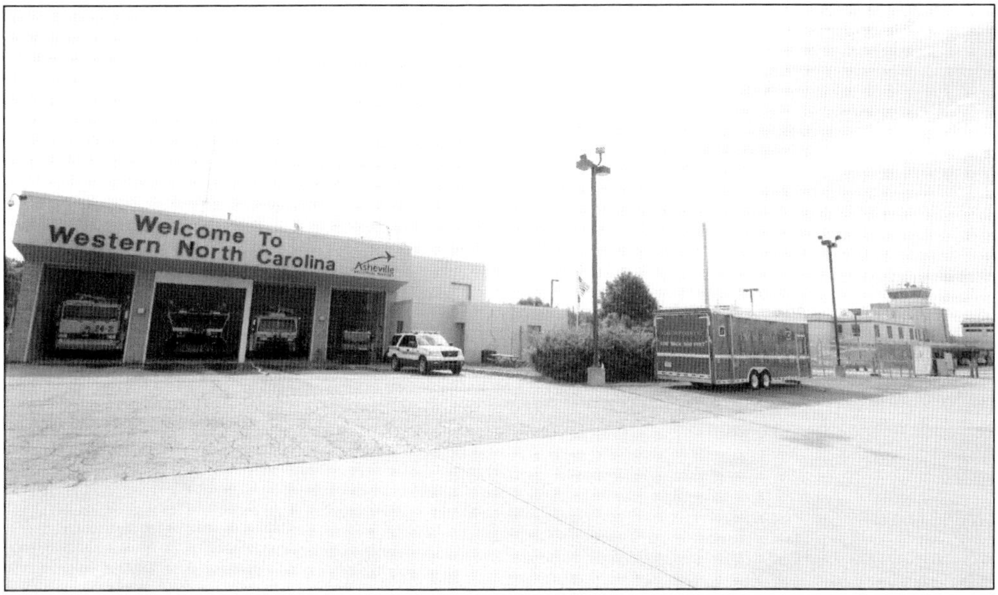

Before this improvement, the airport could only allow five 727s a day. In 1978, the first true four-bay station was built. From 1979 to 1981, DPS operations were suspended, and AFD assumed fire protection with firefighters and a 1961 Ford F750. (Courtesy of Asheville Regional Airport.)

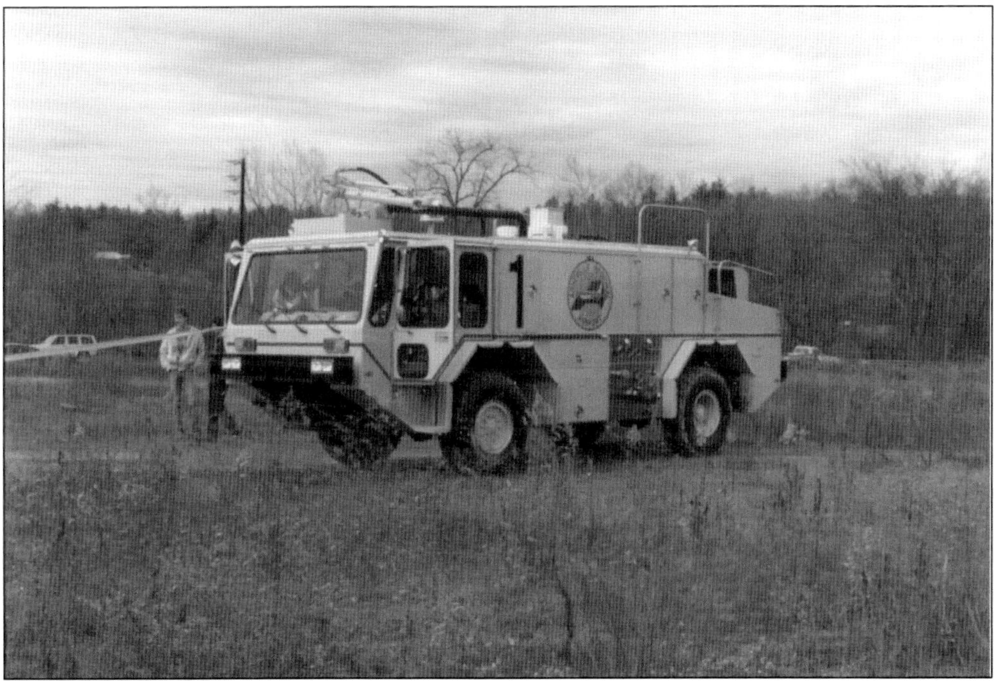

Above, a new E-One twin-agent rapid-intervention vehicle with 500 gallons of water and 500 pounds of dry chemical agent arrives in 1985. In 1988, the name Crash Fire Rescue was changed to Aircraft Rescue Fire Fighting (ARFF). In 1989, a new Oshkosh T1500 twin-agent ARFF vehicle arrived, with 1,500 gallons of water and 500 pounds of Halon. The fire station was expanded in 1993 to include a new communication center, training room, office, sleeping quarters, and day room. (Courtesy of Skyland Fire Department.)

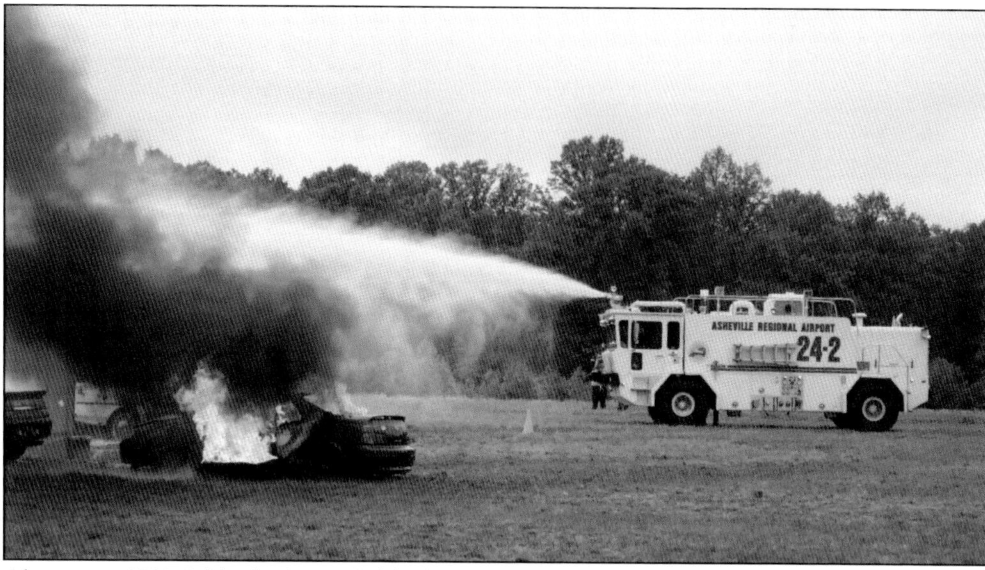

Above is an 1989 Oshkosh. A new Oshkosh Striker 1500 twin-agent ARFF vehicle arrived in 2008. Other unique responsibilities of the airport fire department have included hosting the president of the United States and Air Force One and providing a spray arch from the ARFF trucks for new airlines' first Asheville flights to pass under. (Courtesy of Jeff Augram.)

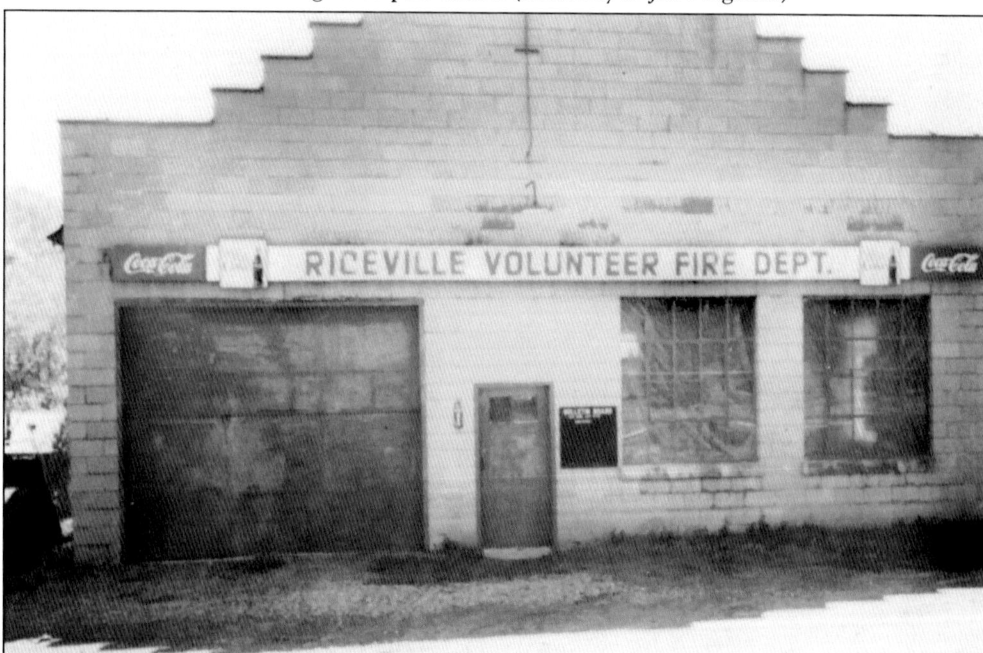

On September 22, 1964, a meeting was held by the Riceville Men's Club at the community center to discuss a fire department. The Riceville Volunteer Fire Department was chartered on October 1, 1964. G.R. Robertson was the first chief. Until February 1965, a person could become a member after attending three consecutive meetings at the community center. At that time, the above building was rented for the first fire station. (Courtesy of Riceville Fire Department and Roy and Polly Armstrong.)

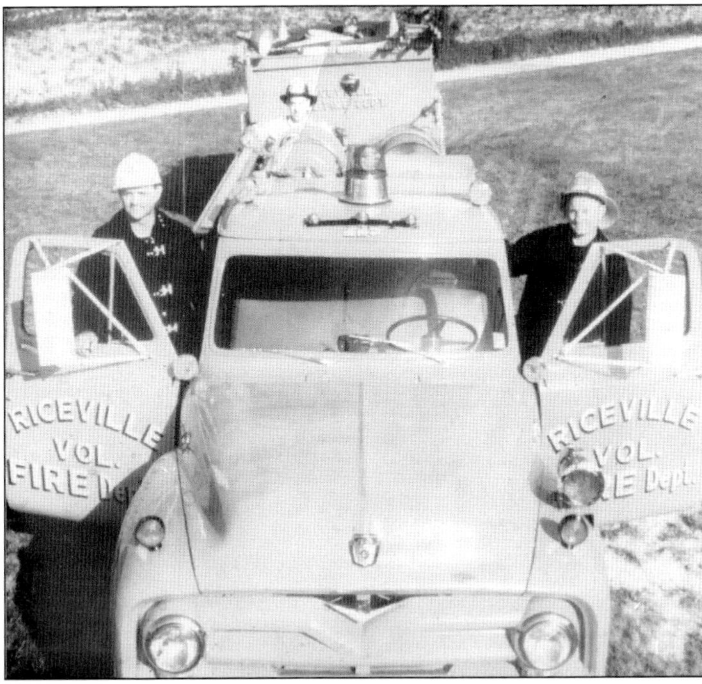

On March 13, 1965, the first truck was purchased for $1,200 from Union Hill Volunteer Fire Department. This truck stayed in service until April 1971, when a crew flew to the American LaFrance Company in Allentown, Pennsylvania, to bring back the new pumper. When they arrived home, phones rang, sirens wailed, and the station's air horn blew to let everyone know the truck they had dreamed about for six years had arrived. (Courtesy of Riceville Fire Department.)

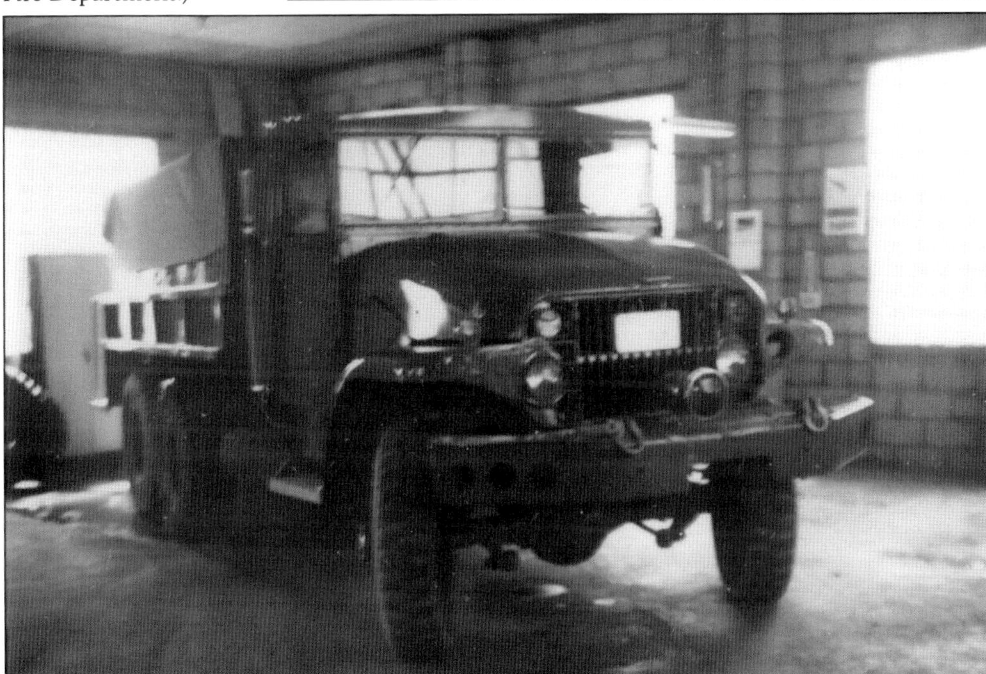

In 1966, an Army surplus six-by-six was purchased for $250 and converted by the fire department into a 2,200-gallon tanker, seen here inside the first station. In November 1970, a search was held for missing citizen Luther Chapman that involved 17 fire departments, the Buncombe County Rescue Squad, a helicopter from Fort Bragg, and famous search dog King and his trainer, Tom McGuinn, who were flown in by Billy Graham from Philadelphia, Pennsylvania. (Courtesy of Riceville Fire Department and Roy and Polly Armstrong.)

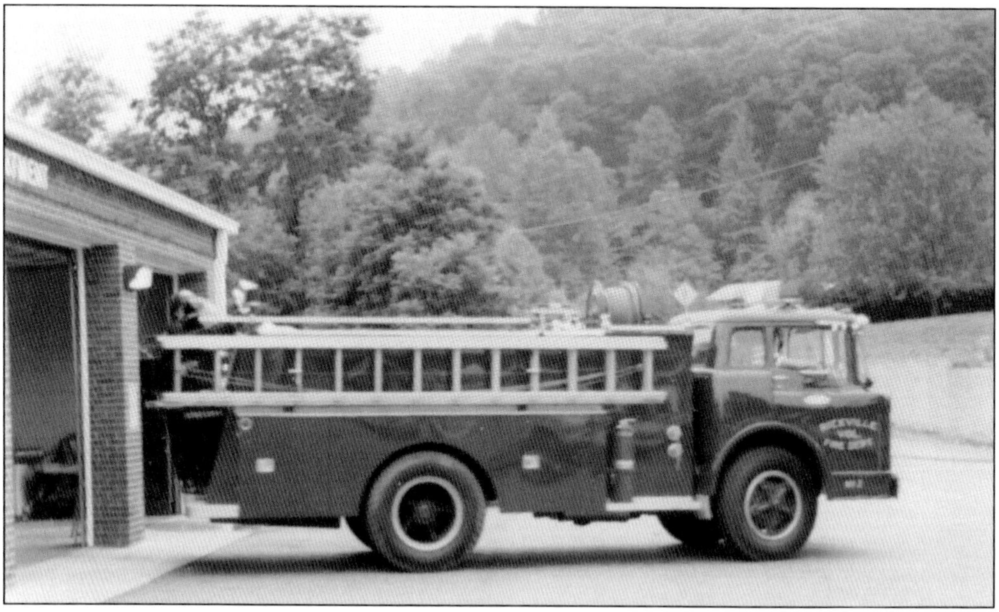

In June 1972, plans were drawn, a loan was obtained, and a contract was awarded for a new station on three acres of land that had been bought on November 20, 1965, and transferred to the Riceville Community Center. Above is a classic photograph showing two firefighters riding the back tailboard, a common practice until the late 1980s. (Courtesy of Riceville Fire Department.)

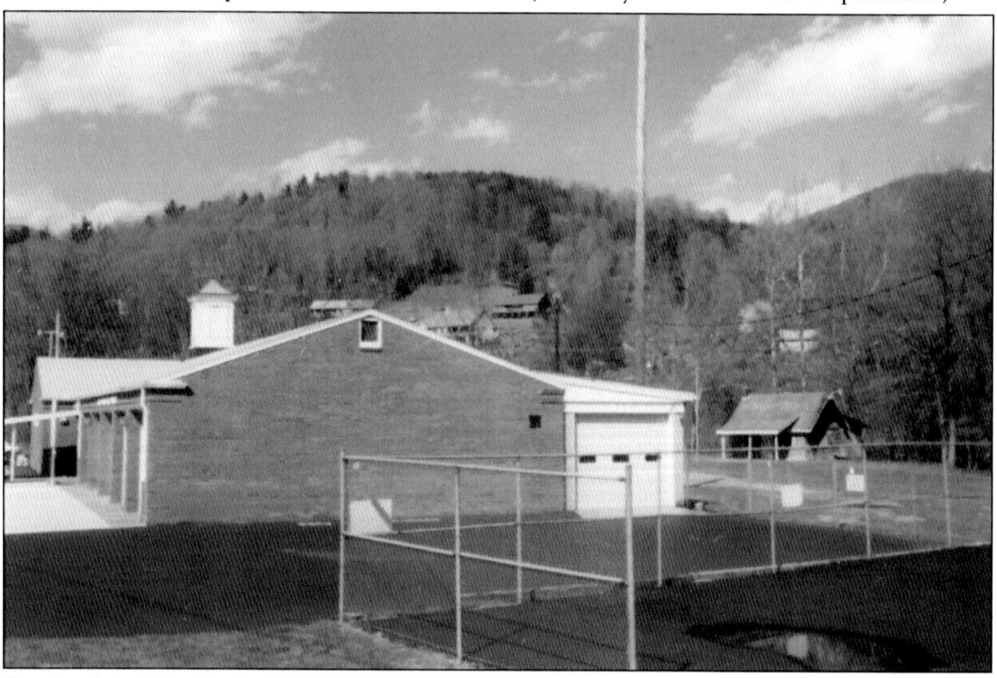

On April 6, 1973, the new station and community center were accepted and final payment was made. An open house was held in July. A second addition was built onto the back of the station in 1979 and included a bay and meeting room. A third, two-story addition was added on this end of the station in 1998 that included dorms, a dayroom, a kitchen, and offices. (Courtesy of Riceville Fire Department.)

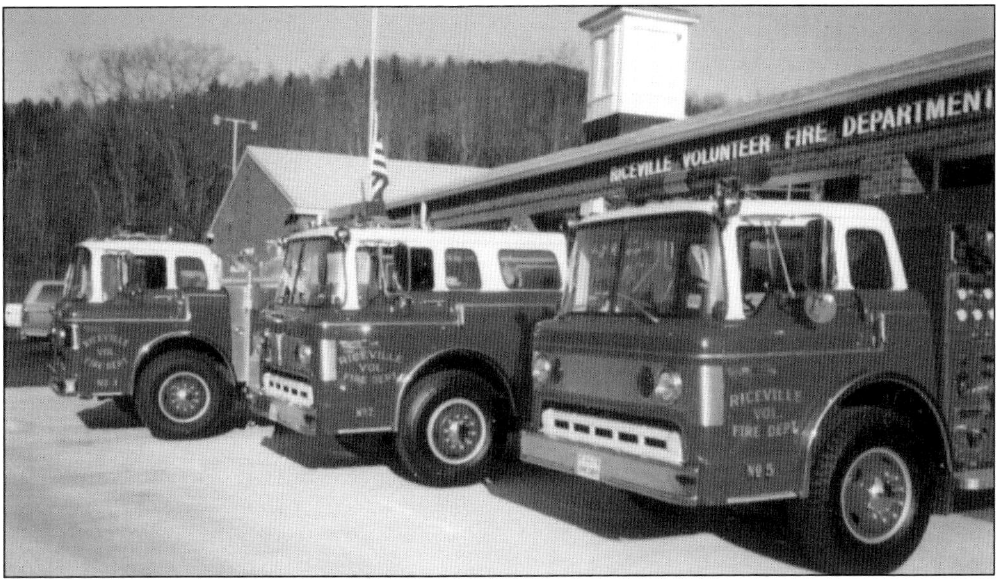

The 1,000-gallon-per-minute Ford Boyer at center arrived in August 1984 costing $105,000. The Boyer truck on the right arrived in March 1986, and the Emergency Equipment, Inc., truck on the left arrived in 1991. The first paid firefighter started in 1987. In 1965, a 1953 Cadillac ambulance was purchased, starting the ambulance service that is still in operation today. (Courtesy of Riceville Fire Department.)

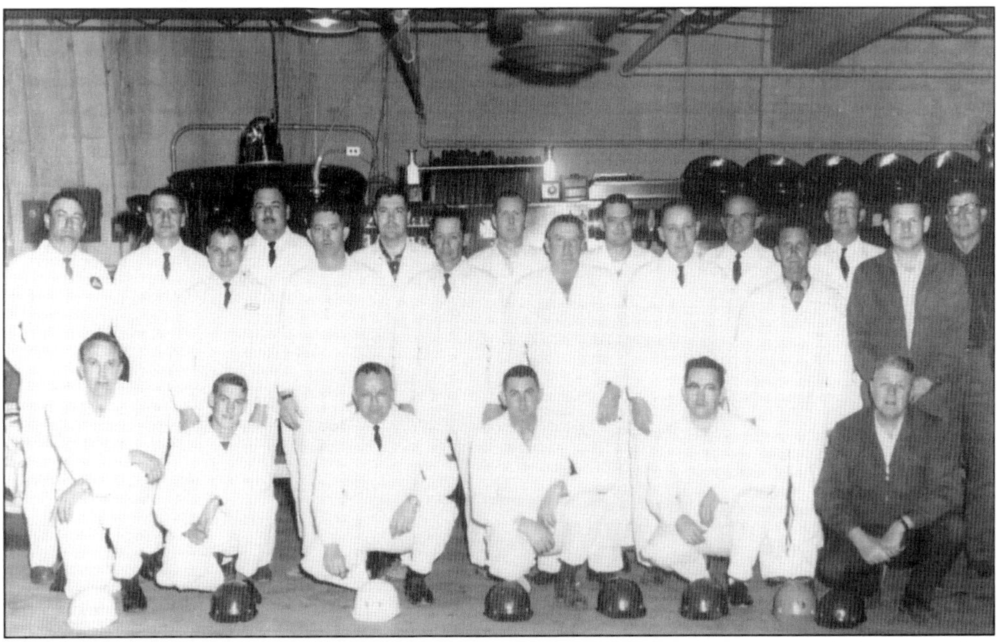

On June 4, 1965, members of the Leicester Grange formed a committee to study forming a fire department and agreed to pay fees for chartering. On July 8, 1965, the Leicester Volunteer Fire Department was chartered. Stan Sluder was elected first chief. They first used a church parsonage. In March 1966, the Buncombe County Board of Education agreed to lease the unoccupied vocational agriculture building at the Leicester School for a fire station. (Courtesy of Leicester Fire Department.)

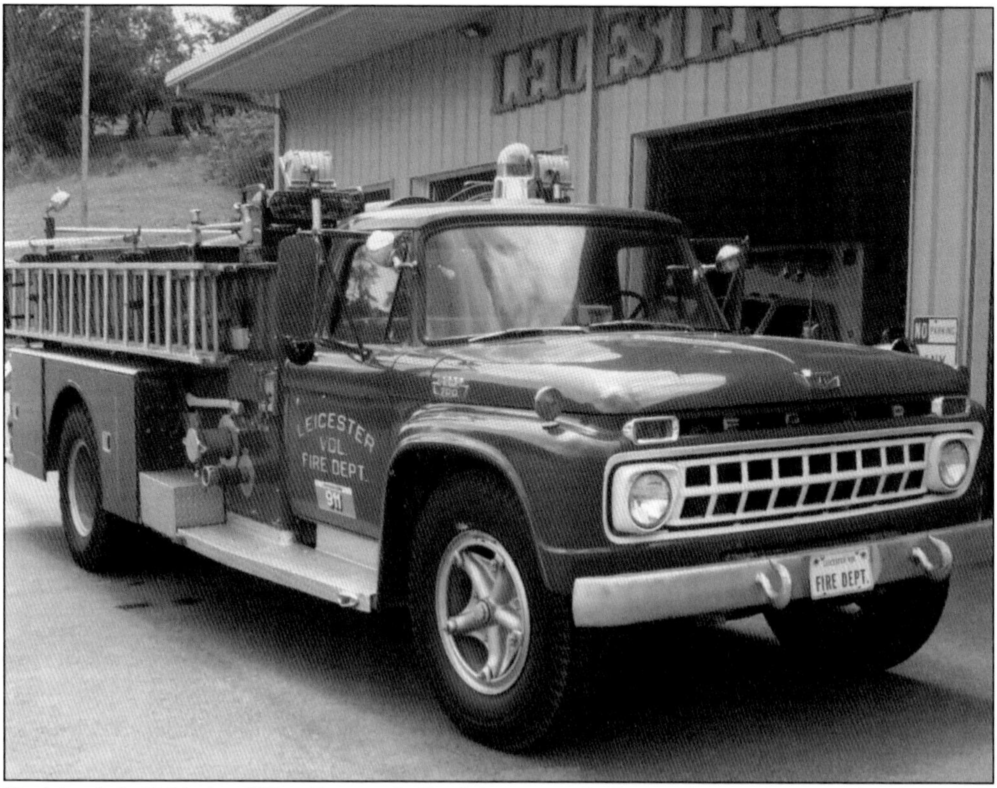

On March 8, 1966, this 500-gallon tank, Ford American LaFrance fire truck was ordered at a cost of $16,090. In 1966, ambulance service started with a Blue Goose Cadillac from a funeral home. Also that year, a 1957 Pontiac ambulance was purchased for $350 from the Buncombe County Rescue Squad. On November 2, 1968, the first of several turkey shoots was held to raise funds. (Courtesy of Leicester Fire Department.)

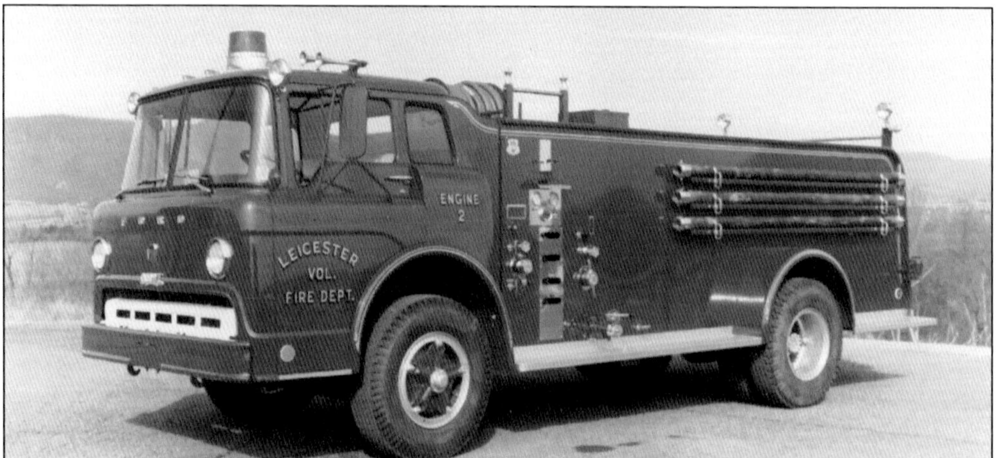

In January 1973, the fire department became the first volunteer ambulance service with a mobile cardiac-care unit and two cardiac medications. A contract was signed for $13,000 on June 28, 1973, with Howe to place a 1,250-gallon tank on a 1968 Ford chassis that the fire department had already bought. That truck is shown above. (Courtesy of Reynolds Fire Department.)

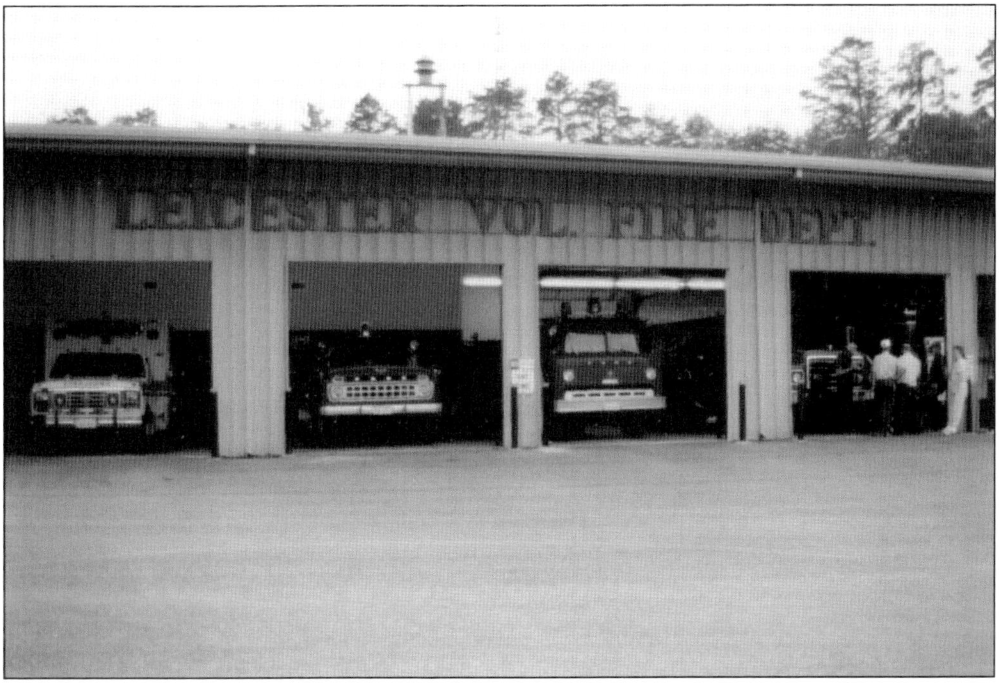

Land for the above station was approved for purchase on April 18, 1975. On February 22, 1976, Leicester suffered a terrible disaster when a dirt-and-rock dam broke at the head of Newfound Creek, killing four members of a family and damaging at least 38 farms. A grand opening was held for the first permanent fire station on October 29, 1978. (Courtesy of Leicester Fire Department.)

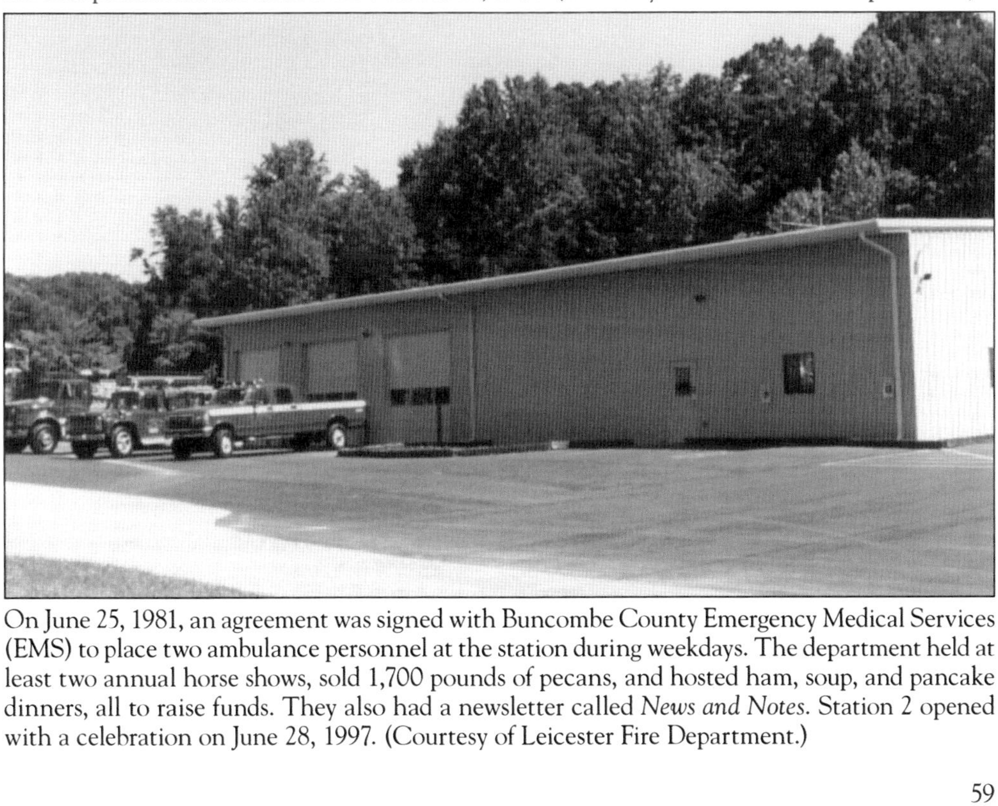

On June 25, 1981, an agreement was signed with Buncombe County Emergency Medical Services (EMS) to place two ambulance personnel at the station during weekdays. The department held at least two annual horse shows, sold 1,700 pounds of pecans, and hosted ham, soup, and pancake dinners, all to raise funds. They also had a newsletter called *News and Notes*. Station 2 opened with a celebration on June 28, 1997. (Courtesy of Leicester Fire Department.)

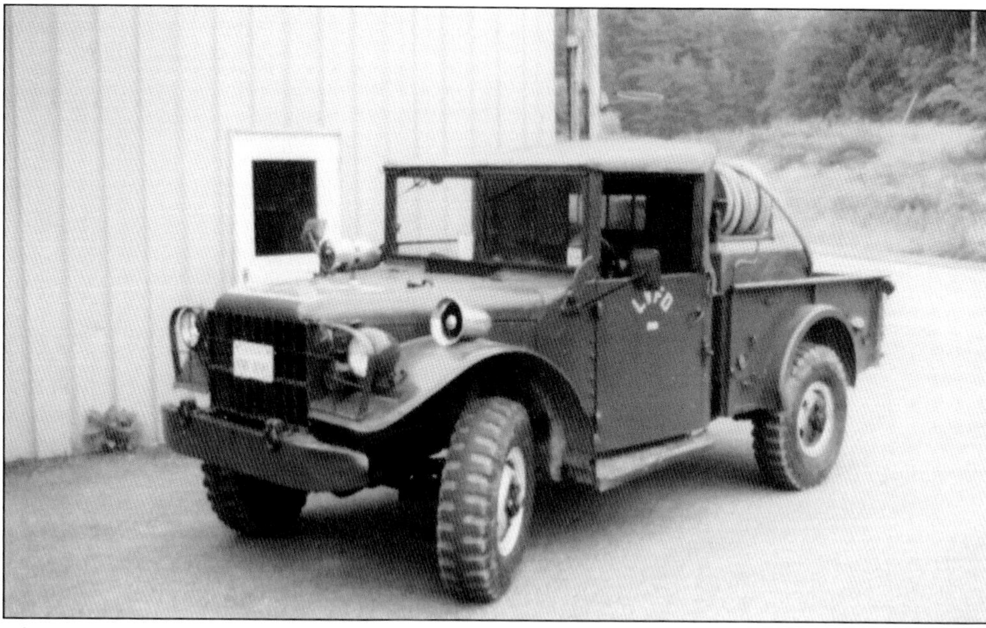

This truck was a surplus military vehicle used as a brush truck, a common practice. A paid firefighter started in the fall of 1995, and Station 3 opened on July 10, 2012. The area got its name from Englishman Leicester Chapman, born 1818 and named for his godfather, the Earl of Leicester. In 1859, he bought 165 acres on Newfound Creek. He also had a store in the area. (Courtesy of Leicester Fire Department.)

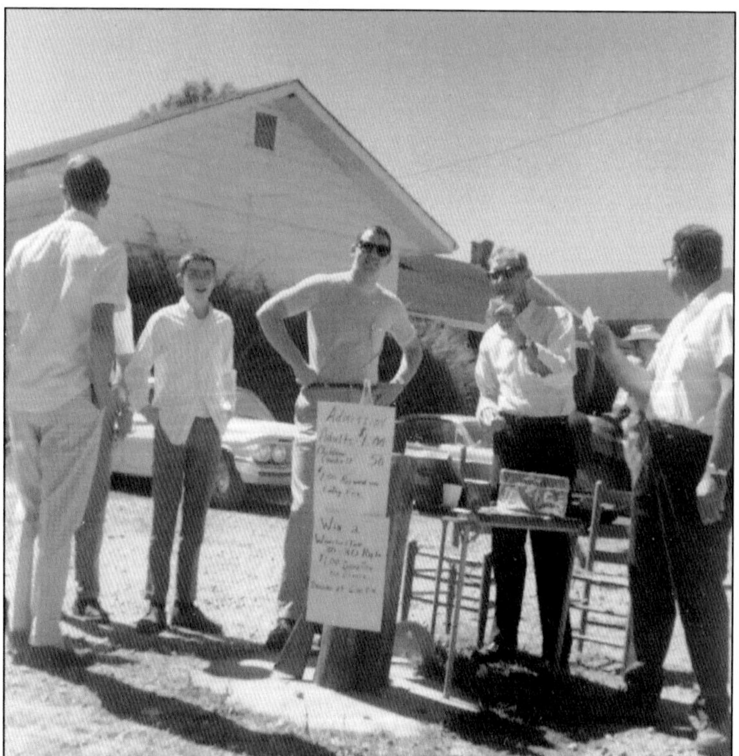

During the first week of May 1966, a group of citizens got together and formed the Reems Creek Valley Fire Department to protect the communities of Beech, Hemphill, Ox Creek, and Pleasant Gardens. Bob Wallen was the first chief. The department made plans for a horse show at the Beech community center on June 4 to raise funds. The first meetings were held at the Ox Creek community center. (Courtesy of Reems Creek Fire Department.)

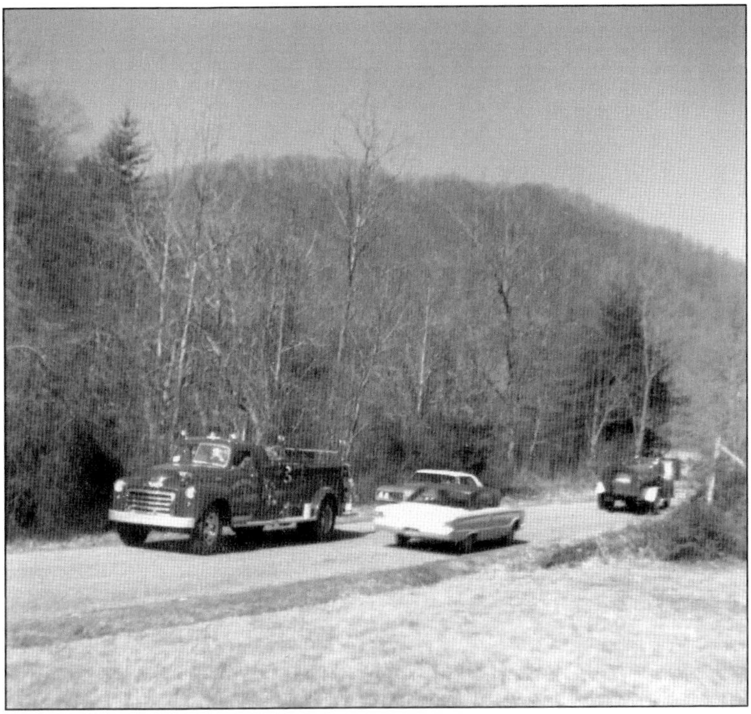

During the first half of September, the group bought 1.5 acres of land for a fire station at a price of $1,125. At the September 22 meeting, it was announced they had bought a 1947 GMC truck, pictured above, for $4,492.90. The back of Jupiter's tanker can be seen driving away. To raise more funds, a pie supper was held, and fire extinguishers were sold. (Courtesy of Reems Creek Fire Department.)

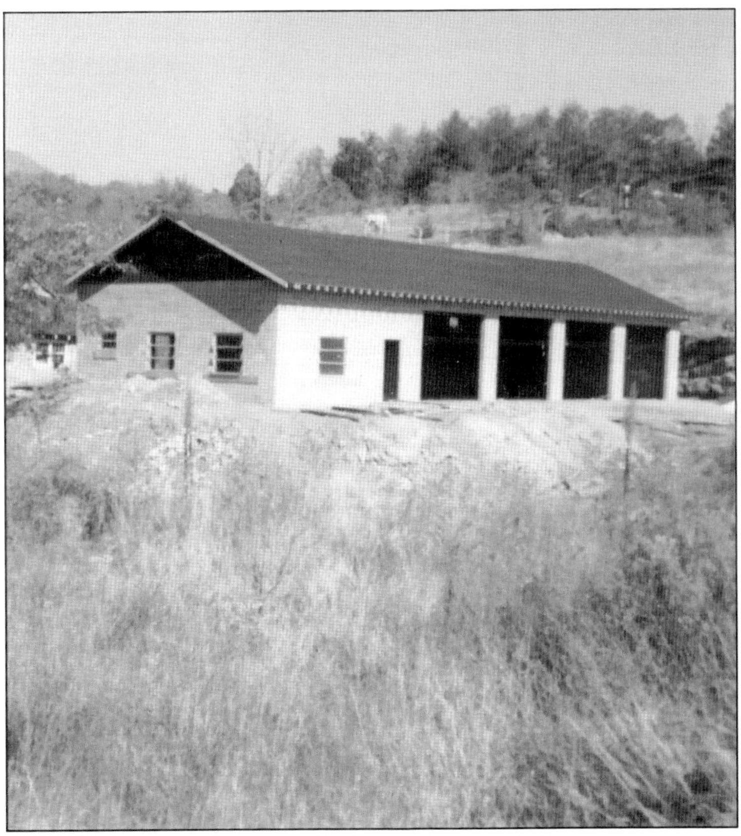

Construction started on the fire station in October 1966. Around November 4, the department received their first alarm at 1:00 a.m. for a fire at a vacant five-room house on Moody Cove Road. Eight men and the chief used their new 500-gallon pumper and had the fire out in five minutes. The house was a loss, but they were able to save an adjacent barn filled with tobacco. (Courtesy of Reems Creek Fire Department.)

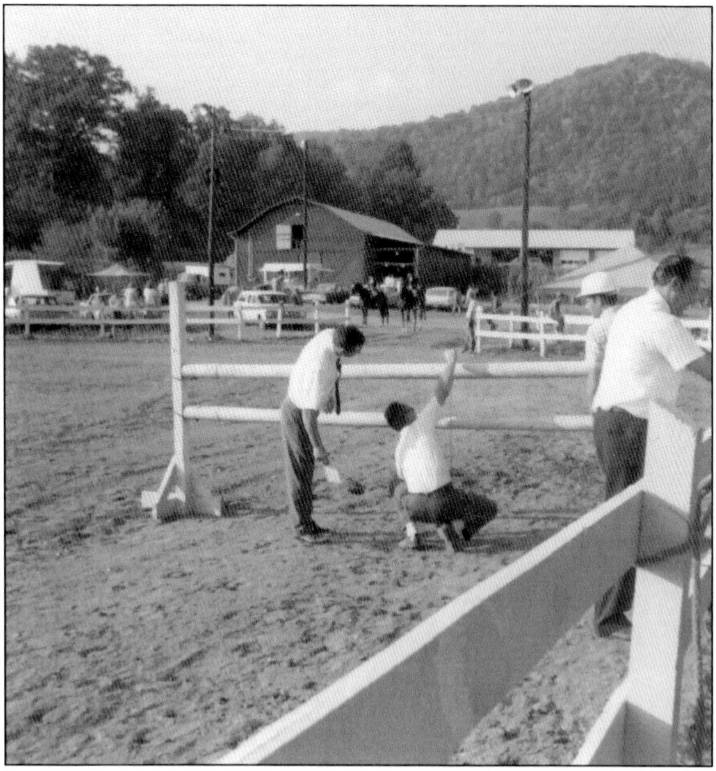

The horse shows moved across from the station (visible in the background of this image) starting with the July 19, 1969 show. Held multiple times a year, these were a major funding source for the fire department with ticket and concession sales. Funds also came from basketball games; suppers; Saturday-night "Fling Dings" with country, bluegrass, and gospel music at the Beech Community Center; and donations. (Courtesy Reems Creek Fire Department.)

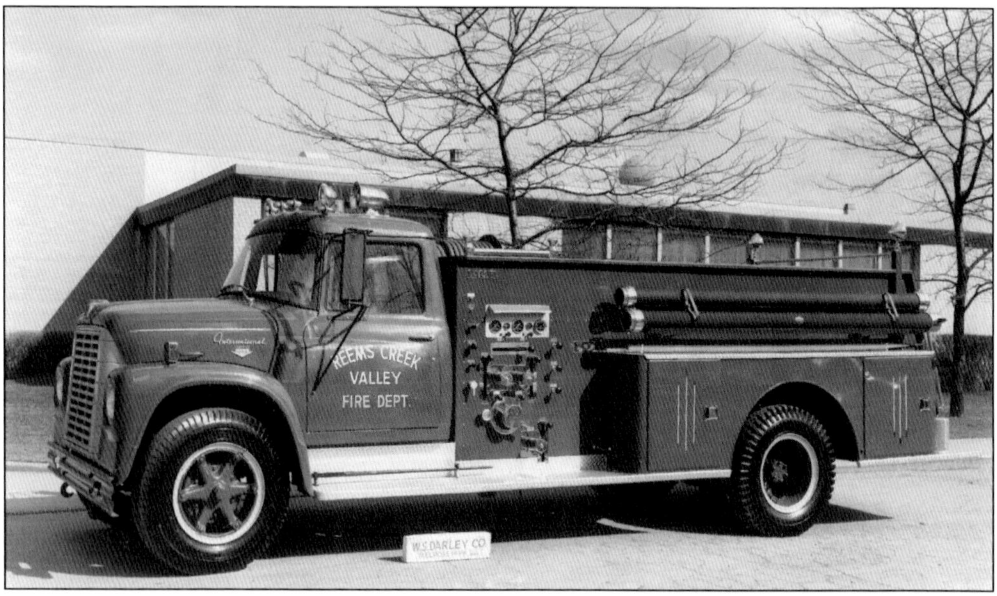

A group of 17 AB-Tech students studying masonry helped build the station, and on January 31, 1967, they announced they would hold a formal dedication on April 16. AFD's Lt. Pat Oxner taught 37 members of the department a month-long training course, and they celebrated their graduation by extinguishing burning oil barrels and old tires. In June 1971, they received their first new International Darley 750-gallon-per-minute truck, which they still own. (Courtesy of Reems Creek Fire Department.)

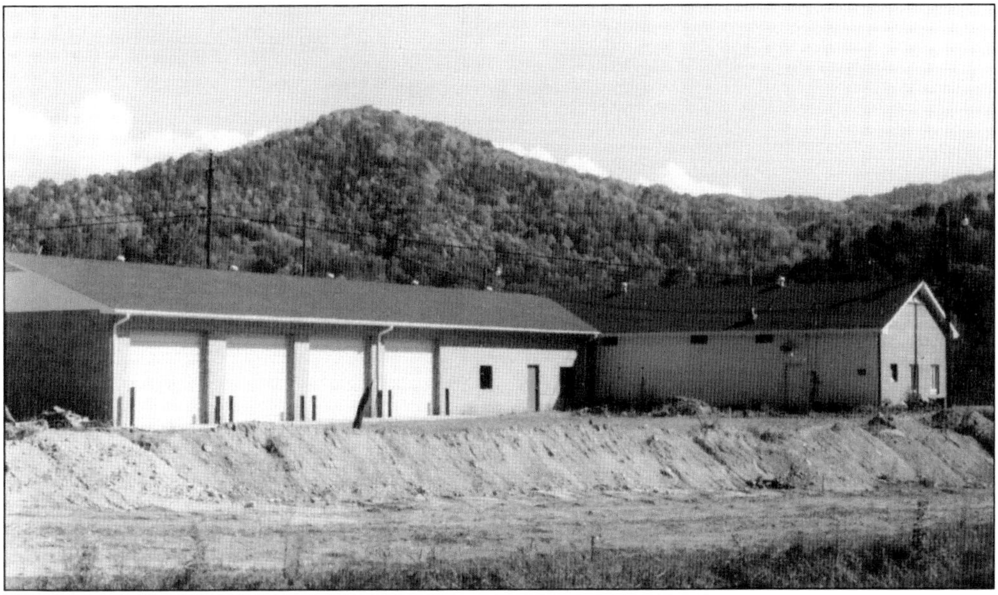

The body of the 1947 truck was moved to a 1975 International. A vote in favor of imposing a fire tax was held on March 27, 1971, and passed. The community made the point that this was the only fire department in North Carolina at the time that did not have a fire tax. A second, four-bay addition was constructed in the 1980s and featured meeting room, a kitchen, offices, and bathrooms. A third, six-bay addition was built onto the second addition and opened in September 2012. (Courtesy of Reems Creek Fire Department.)

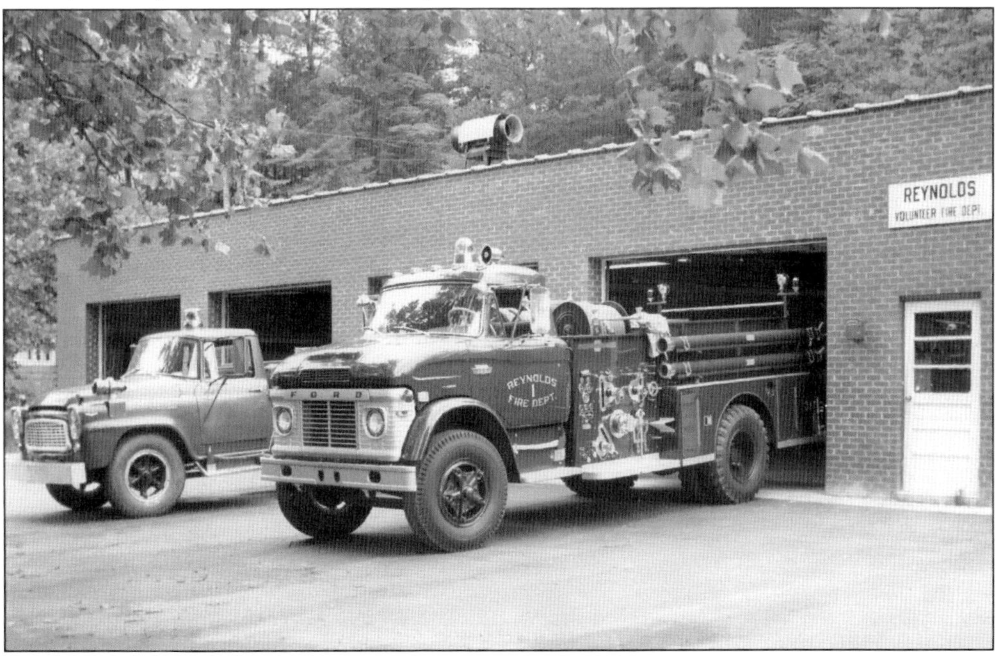

Incorporated on March 11, 1968, the Reynolds Volunteer Fire Department started after a rash of three house fires. Bud Sales was the first chief. One of them burned a major phone line in two that ran right over the top of the house. Their first fire was a brush fire at Recreation Park at the dam with AFD. A fire tax soon passed in November 1968. (Courtesy of Reynolds Fire Department.)

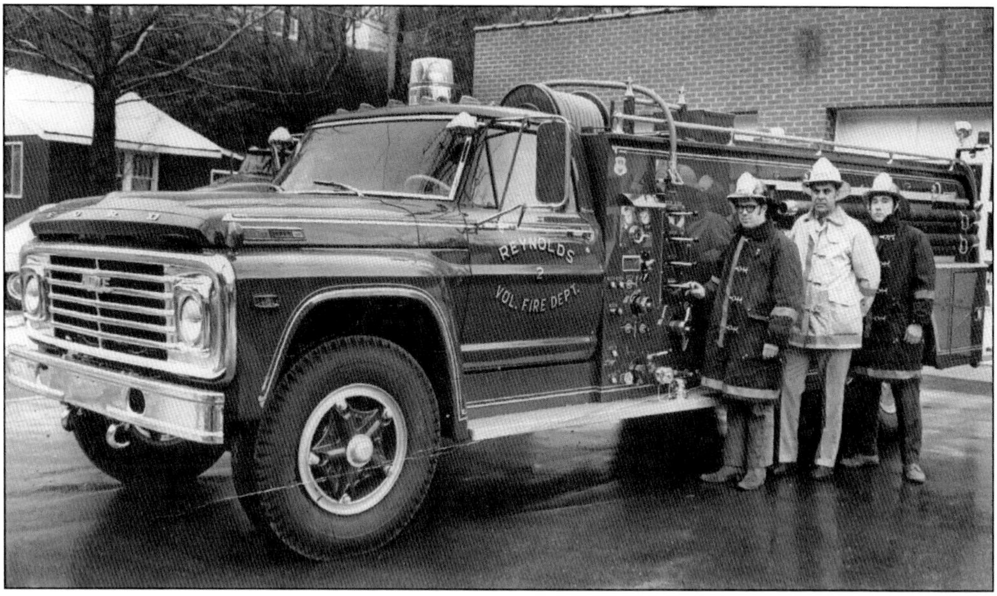

Barnardsville Fire Department loaned the Reynolds Volunteer Fire Department their first truck (the front-mount pump that had also been Barnardsville's first truck), which was kept at Chief Sales's farm. That truck went to French Broad Fire Department next. In the summer of 1968, they bought Barnardsville's 1,000-gallon International tanker. The first four-bay station opened in October 1968. They ordered their first truck in 1968, and it arrived in February 1970. (Courtesy of Reynolds Fire Department.)

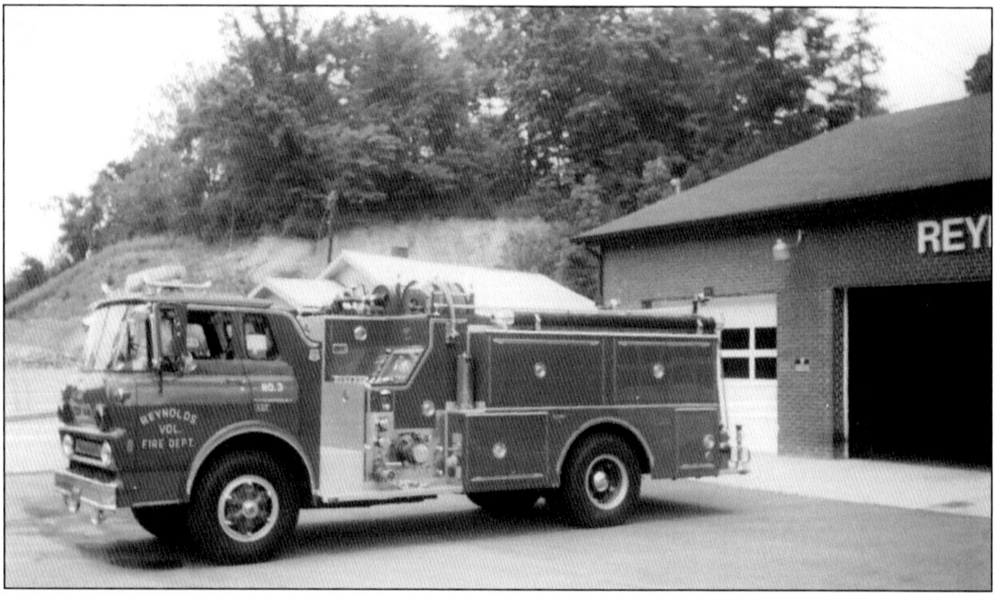

The first new truck was a $20,000 Ford American LaFrance with a 750-gallon-per-minute pump. In January 1973, the department received their second new truck, a $22,000 Ford 1,000-gallon, 350-gallon-per-minute pumper-tanker. It replaced the International tanker and was used by Garren Creek and Upper Hominy Fire Departments. Their third truck was the Chevrolet Howe seen above. In the mid-1980s, a pitched roof and an addition off the back of the bays were constructed. (Courtesy of Reynolds Fire Department.)

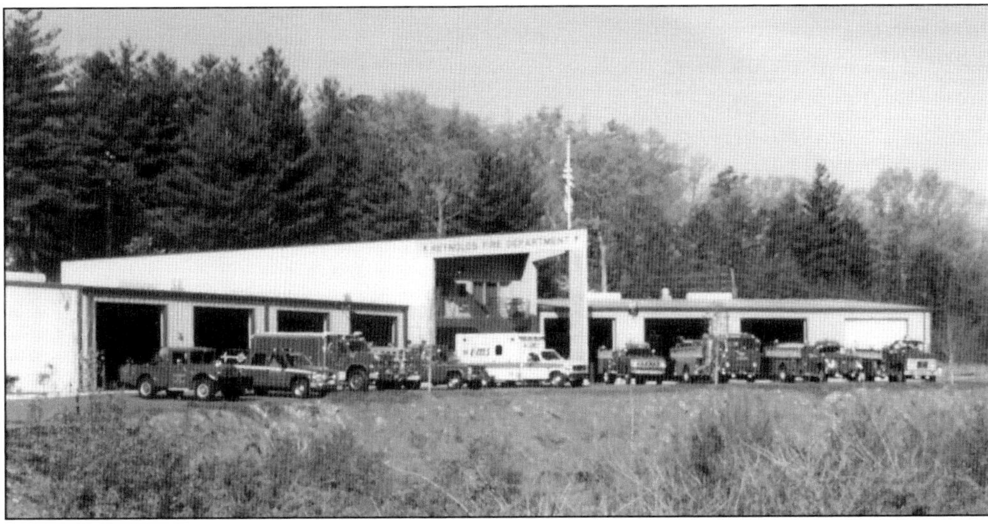

In June 1993, the department moved into its new multimillion-dollar station. The first paid firefighters started in 1995. They used to raise money by selling chocolate, vanilla, and fresh peach and banana ice cream for 25¢ across the street from the station at the Sales family produce stand. At one time, they had 8–10 churns in the station for three months making ice cream. (Courtesy of Reynolds Fire Department.)

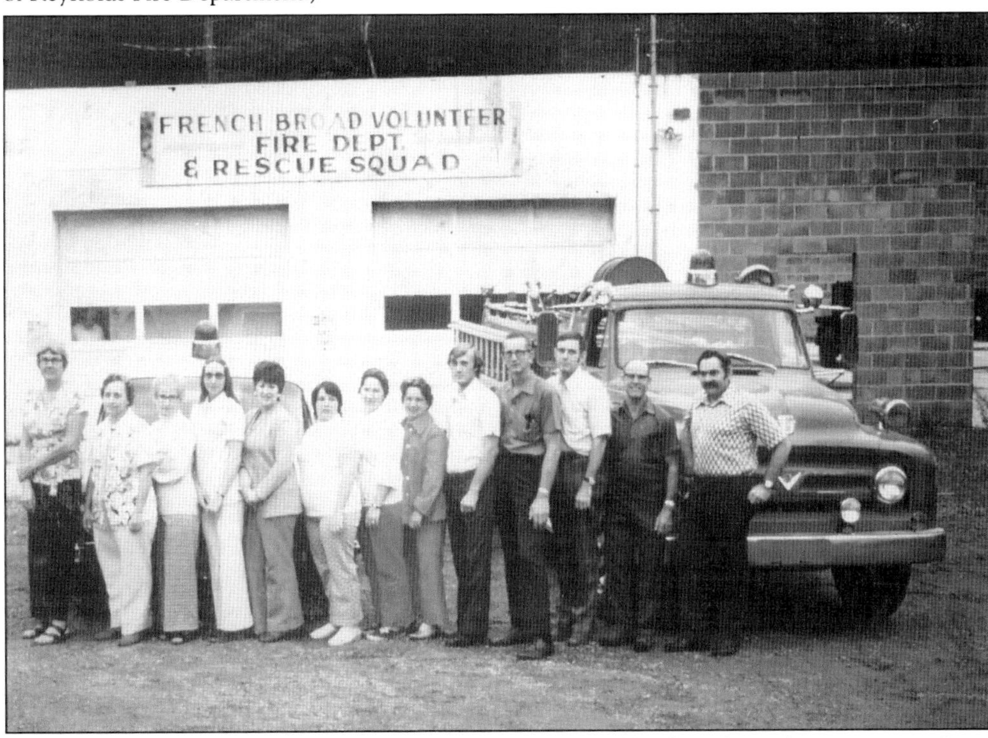

French Broad Volunteer Fire Department (VFD) was started by the Alexander community Lions Club and chartered on January 10, 1969. Leo Snelson was the first fire chief. The foundation for the first station was made, but construction stalled due to lack of funds. The Ladies Auxiliary raised funds to complete the first fire station, shown at left, which included two bays and an office. (Courtesy of the French Broad VFD.)

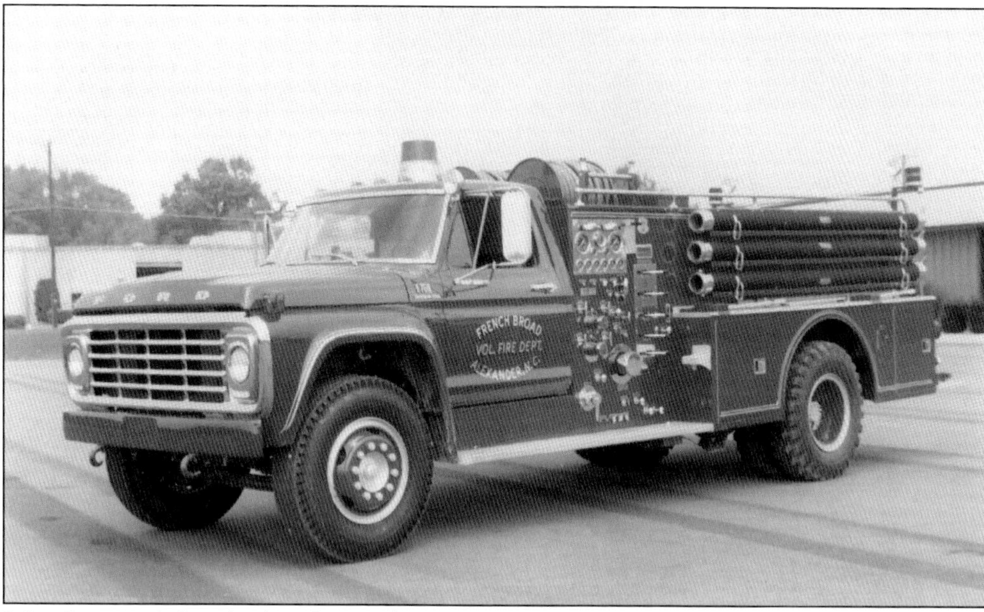

It has been said that the first fire truck was borrowed from Barnardsville Fire Department until enough money was raised to buy a new one. That first truck bought came from the department in Aiken, South Carolina. French Broad's first brand-new truck was this 1976 Ford Howe. The department also started with a hearse ambulance service. (Courtesy of the French Broad VFD.)

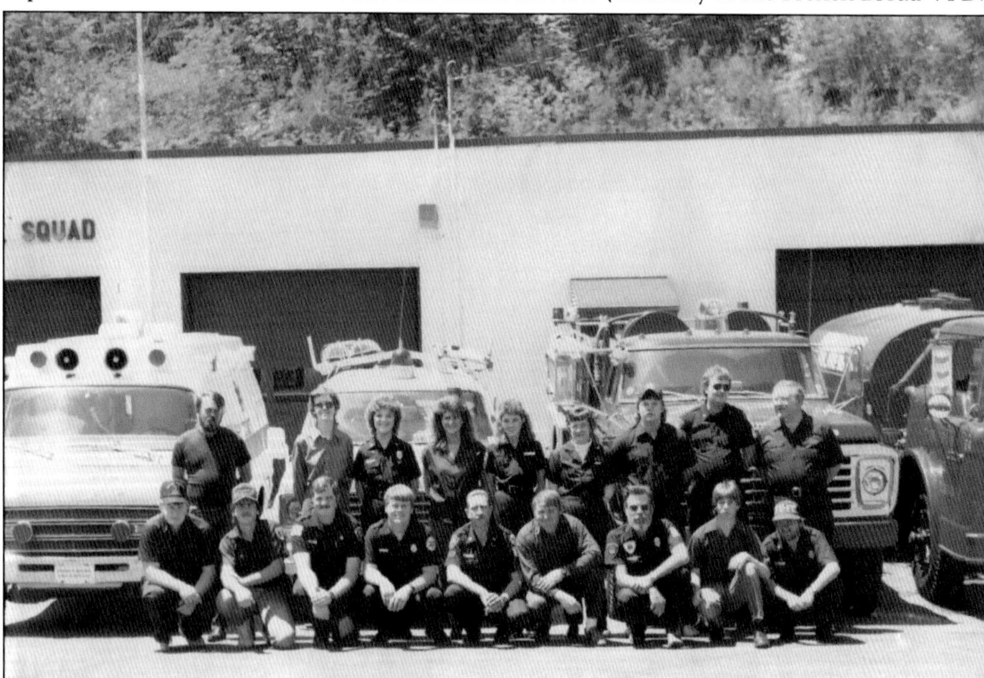

Members bought station supplies on their own, but money was also raised to run the fire department through cakes sales, dinners, door-to-door solicitation, rummage sales, cookouts with live bluegrass music at the firehouse, and the use of the fire truck to pull cars up snowy hills in the winter. (Courtesy of the French Broad VFD.)

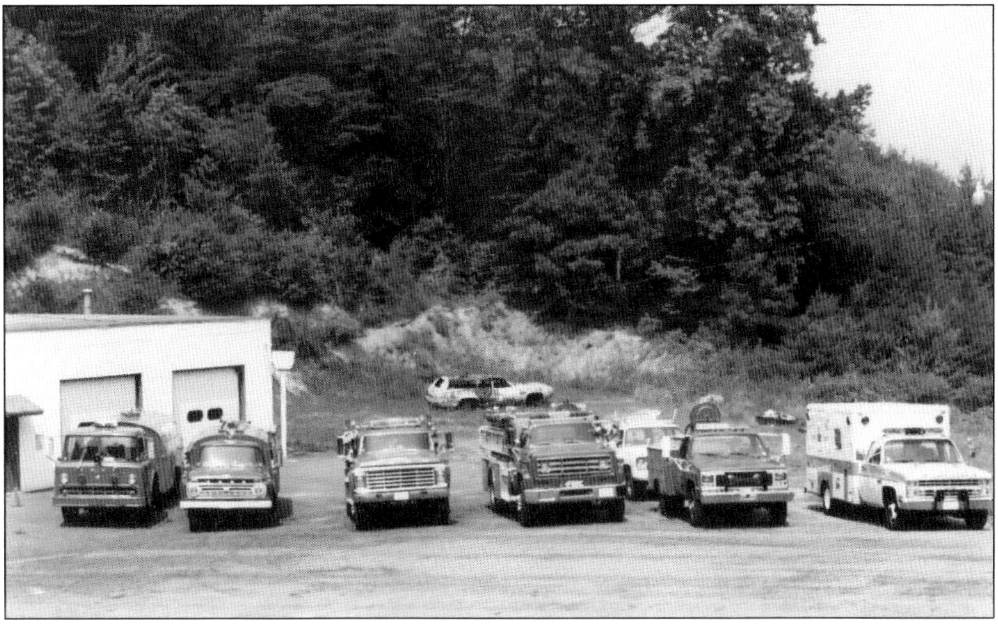

Money was raised to build the second addition, shown above, by selling concrete blocks for 25¢ apiece. Sometime in the mid-1970s, a Farmers Federation Loan was approved for $50,000 to complete the fire station and purchase equipment. This addition was inspected and approved in October 1975. In 1991, a third, four-bay addition was built on. Paid daytime personnel started in June 1998. (Courtesy of the French Broad VFD.)

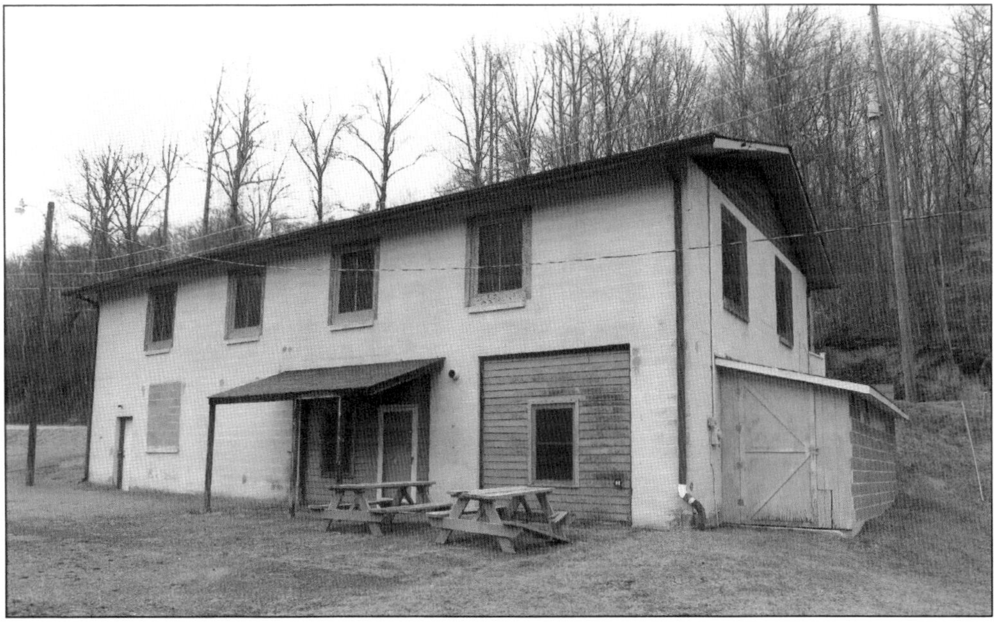

Garren Creek Fire Department was originally charted as Flat Creek VFD on September 19, 1977. The name was changed on May 23, 1978, after they discovered another fire department had the same name. Their first truck was a 1964 Chevy American LaFrance with a 750-gallon-per-minute pump. Their first station was in the basement of the already-present community center, shown above. (Author's collection.)

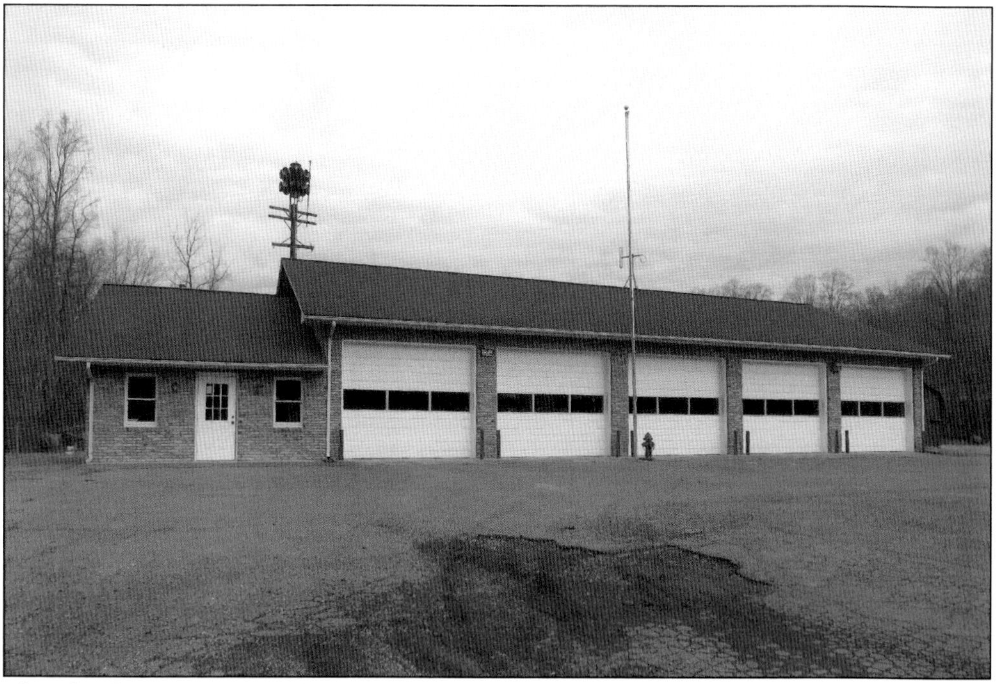

Their first built fire station, shown above, came in 1988. This station became a substation when their current station was erected in 2008. (Author's collection.)

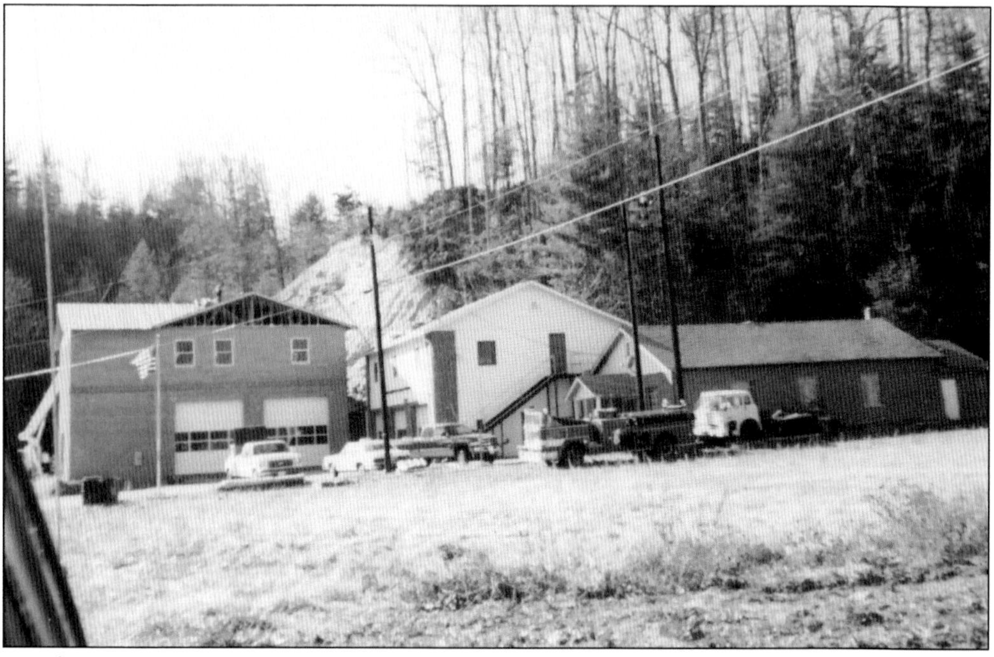

Broad River Volunteer Fire Department was chartered on January 20, 1978. The department built its own first truck. A two-bay station was built onto the 1950s community center, which had a dirt floor, until the fire station was added on. In the late 1980s, a two-story, three-bay, second addition was added onto the front and top of the original two bays. This addition was not completed until 1994. (Courtesy of Broad River Fire Department.)

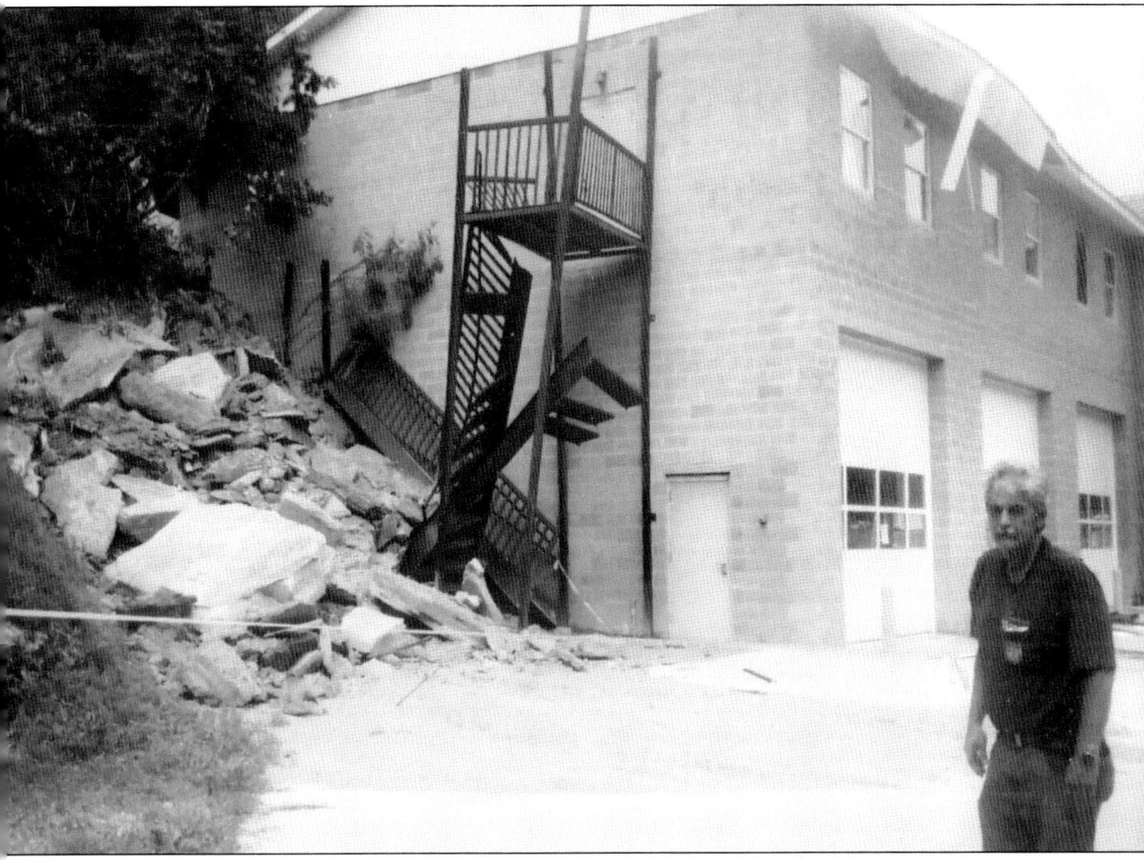

On June 25, 2005, as the third addition was nearing completion, a rock slide occurred on the slope cut back to build the addition. A second slide occurred two days later. Repairs were made, and a dedication was held on December 25, 2006. The Schumont substation was built in 2005 from land and funds from a single benefactor. Paid firefighters were hired on July 1, 2007, the last in Buncombe County. (Courtesy of Broad River Fire Department.)

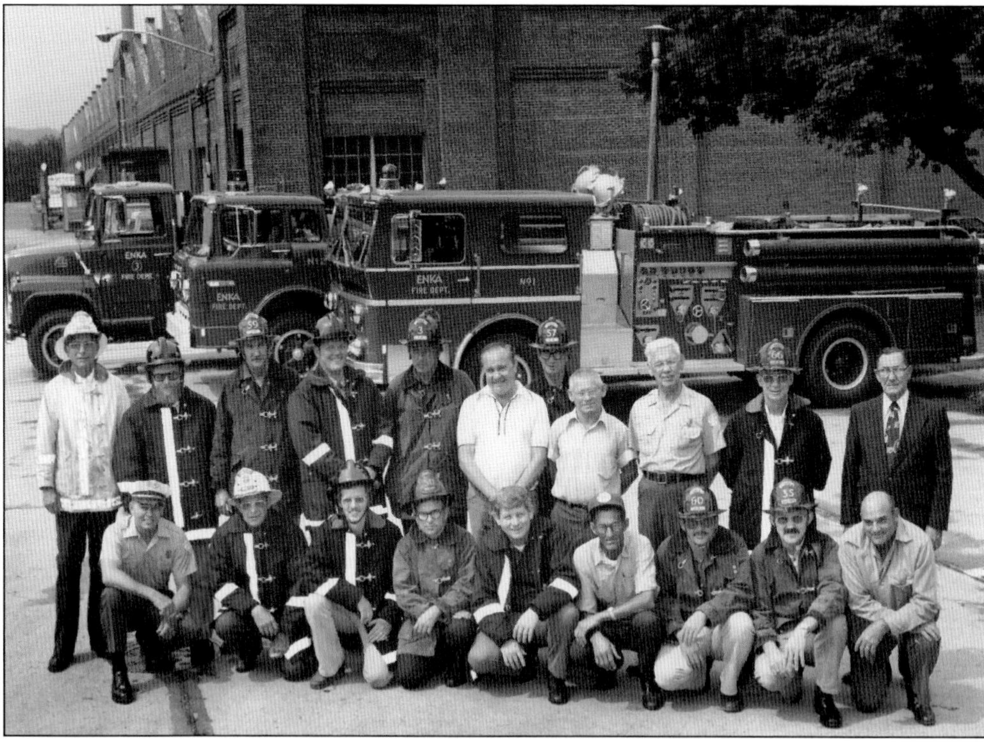

In late 1986, BASF announced it could no longer answer calls outside of the plant. BASF donated $40,000, the above cab-over Ford 500-gallon-per-minute pumper and Ford 1,000-gallon tanker, a station wagon, and various equipment for a new fire department. Enka-Candler Fire & Rescue Department was formed and chartered on February 9, 1987, with Jim Cathcart as the first chief. (Courtesy of Enka-Candler Fire & Rescue.)

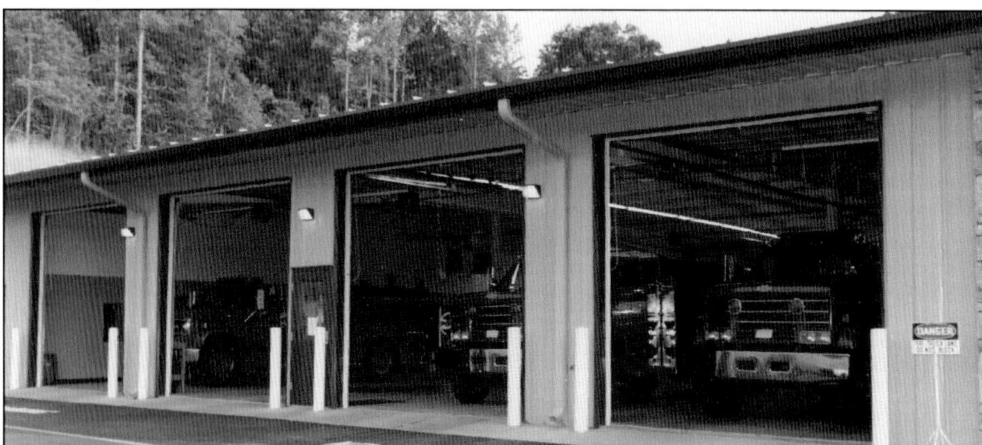

Enka-Candler Fire & Rescue started answering calls on January 1, 1988, out of the BASF plant fire station while their facility was being built. They moved in mid-March, and on May 28, a dedication ceremony was held. An amount of $800,000 was spent on the new station and three additional pumper-tankers, shown above. In October 1988, the first paid firefighters were hired. An office addition was built around 2003. In February 2005, Station 2 opened. (Courtesy of Enka-Candler Fire & Rescue.)

Three

FORMER FIRE DEPARTMENTS

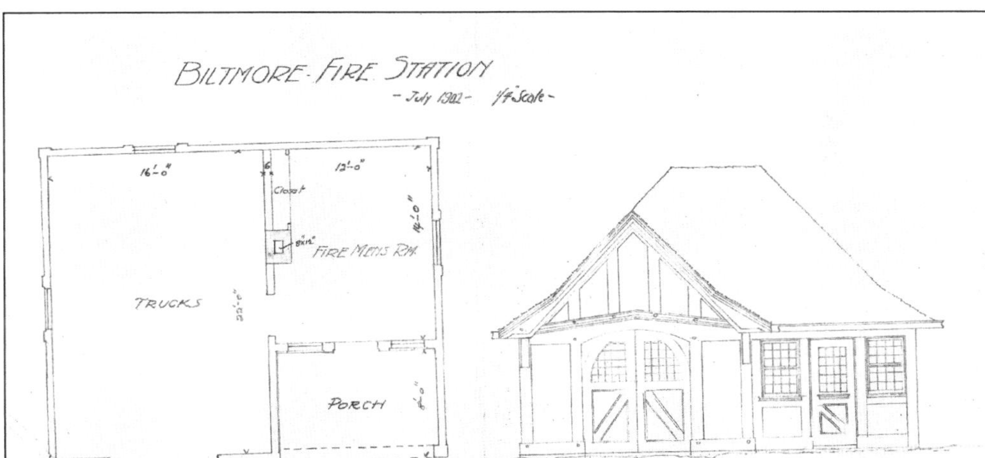

The Biltmore Fire Department was started on April 6, 1906. J.C. Lipe was the chief and served up until his death in the great flood of 1916. The above station seems to have been built in 1902 on Brook Street in George Vanderbilt's town of Biltmore, now known as Biltmore Village. Biltmore Volunteer Fire Department existed starting 1903, but not much is known about them, and they may have dissolved prior to 1906. (Courtesy of Collection of Asheville Art Museum, Gift of Historic Resources Commission, Asheville and Buncombe County.)

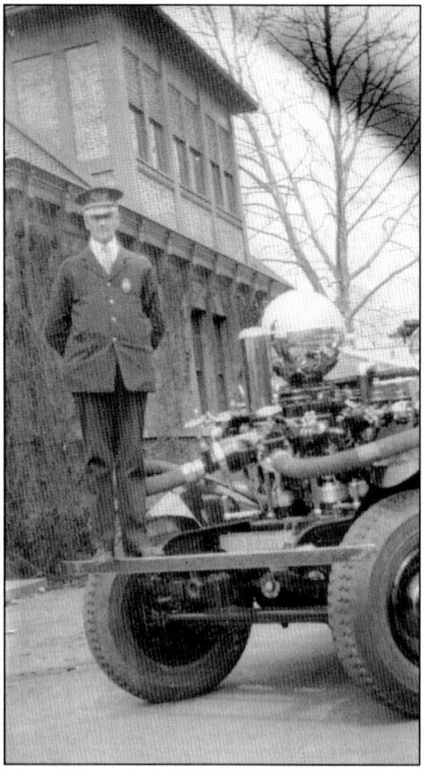

Until at least 1918, the department had one hand-drawn hose wagon and one hose reel with nozzles and hose. They had 10 to 15 volunteers and a fire bell at the station. Edith Vanderbilt sold Biltmore Village at the end of 1919–1920. By 1925, they had an ALF engine. On July 1, 1926, they bought an Ahrens-Fox demonstration truck, which AFD acquired (shown above) when the city annexed the area in 1929. (Courtesy of the James Jackson family.)

Edwin Carrier bought the West Asheville Improvement Company and 1,200 acres in March 1892. The town of West Asheville was eventually incorporated. In December 1914, the alderman authorized the purchase of 1,200 feet of fire hose and two 500-foot hand reels for a volunteer fire department to be organized. A week later, members of Company No. 2 had a meeting presided over by Chief A.M. Young in which they divided their half of the town into four districts and named a captain, a lieutenant, and firemen for each. Company No. 1 did the same. West Asheville became part of Asheville in 1917. (Courtesy of Environmental Data Resources, Inc.)

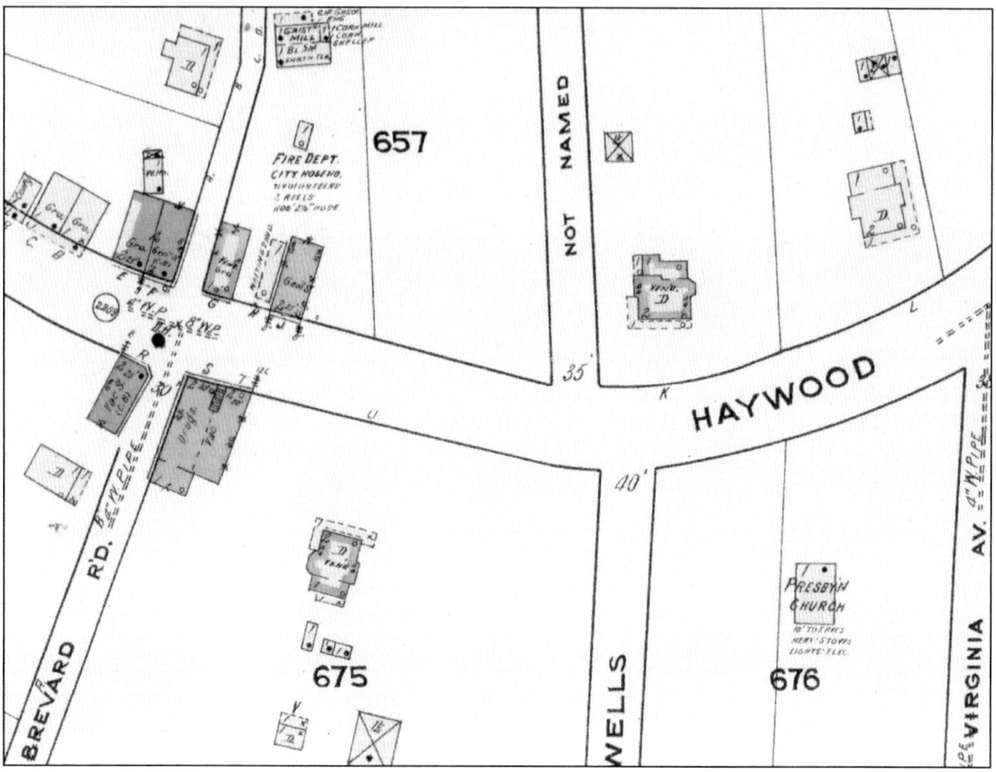

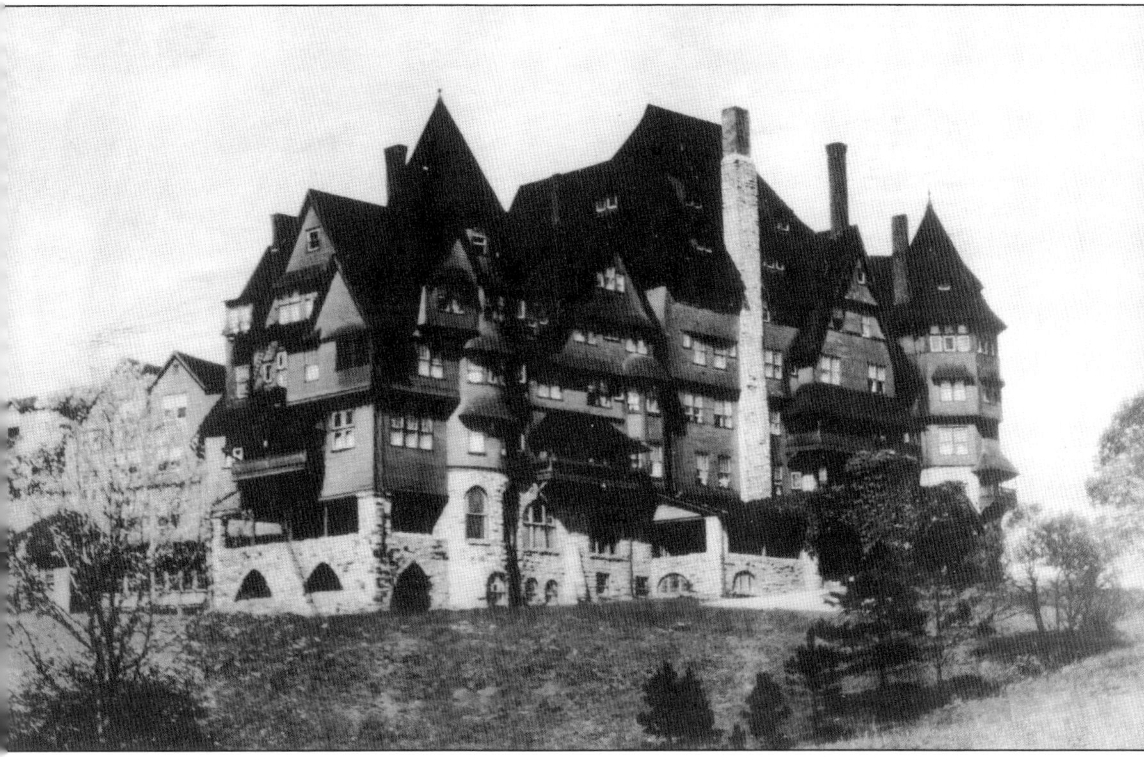

The first Kenilworth Inn was completed in late 1890 to compete with Asheville's other luxury hotels: Battery Park and Belmont. Kenilworth and Battery Park were nationally recognized for their elegance. The owner, former Pennsylvania state senator Joseph Gazzam, formed the unincorporated town of Kenilworth on February 27, 1891. The massive wood-frame building stood above Biltmore Village on a 151-acre park. The inn was completely destroyed by fire on April 14, 1909, and the ruins were bought by James Madison Chiles in 1912. Chiles incorporated the town of Kenilworth in 1913 and became its first mayor. He started rebuilding the inn—this time to compete with the Grove Park Inn, Manor Inn, Langren Hotel, and the Battery Park Hotel—in November 1916, and it was nearly completed in January 1918. The Army started using it in February for General Hospital No. 12 and continued to do so until September 1919. Chiles leased 337 acres in Azalea to the Army to build General Hospital No. 19. When this hospital was fully operational, Kenilworth was deactivated. (Courtesy of North Carolina Collection, Pack Memorial Public Library, Asheville.)

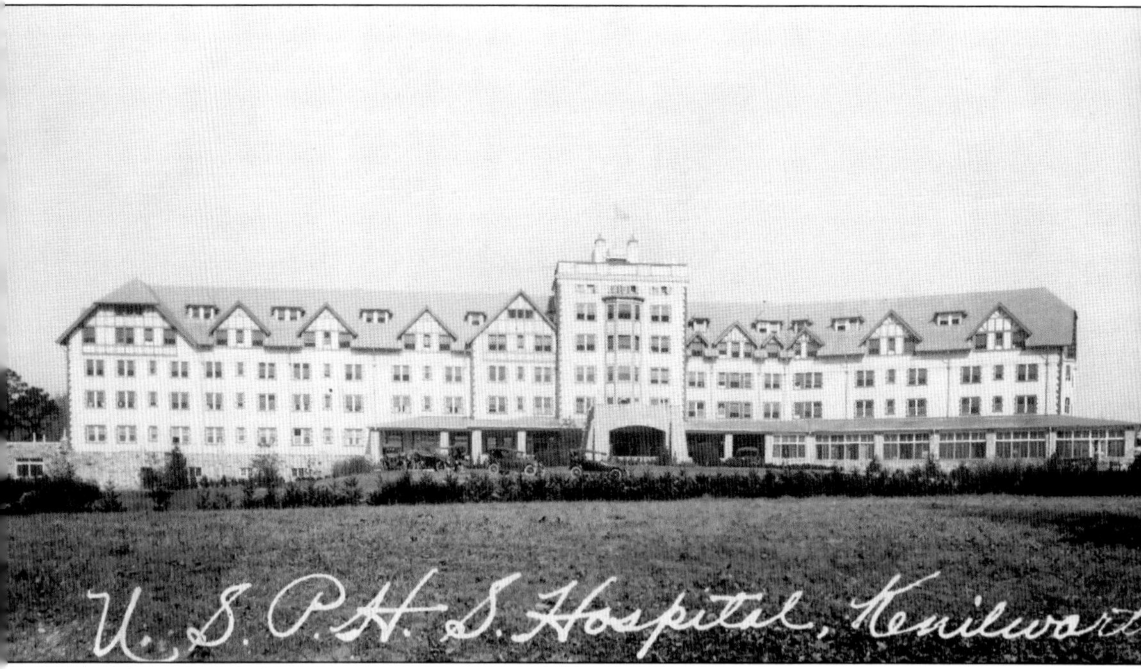

The Army took two years to return the inn to its original state, and the Kenilworth Company took almost another year to add finishing details. The 500-guest inn was finally opened on March 10, 1923. It existed until the banks closed during the Great Depression, causing the company to default on its loan in 1930. It sat empty before becoming the mental health facility Appalachian Hall in October 1931. In February 1943, it went back into military service during World War II as the US Naval Convalescent Hospital. Nearly 7,000 patients were treated there before it was deactivated in April 1946 and converted back into Appalachian Hall. Owners changed, but the name and focus continued until 1994, when the facility changed hands again. It was still a mental hospital but was now called Charter Asheville. Finally, on October 22, 2000, it ceased being a mental hospital and was remodeled for apartments. (Courtesy of Charles George VA Medical Center.)

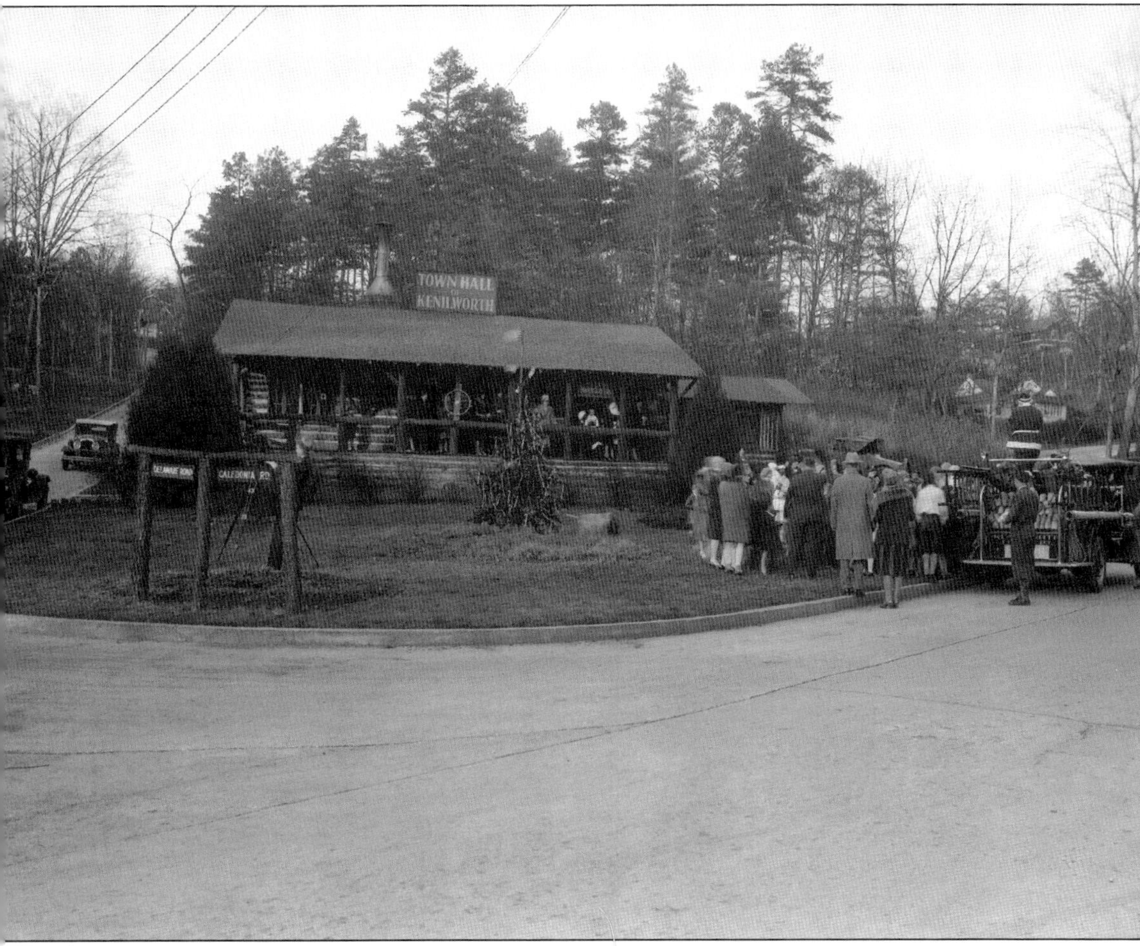

When Kenilworth Inn was being constructed, the plan was to have two 500-foot, 2.5-inch hose reels on the property. This 1921 photograph shows volunteer Kenilworth Fire Department's 1918 (the year the inn was first used) American LaFrance fire truck. This truck was probably the first and only truck they had. Santa Claus is pictured standing on the truck in front of the Kenilworth Town Hall. James Chiles was the first mayor, serving until his death in 1925, and his wife, Leah, became mayor in November 1928, serving until the town dissolved in 1929. The fire station was located in a garage across the street from the rear of the Kenilworth Inn. The 1925 Sanborn Insurance Map for the area stated Kenilworth had a Knox fire truck that year. The town voted to be annexed by Asheville, which occurred on June 30, 1929, ending the fire department. (Courtesy of E.M. Ball Collection, Special Collections, Ramsey Library, UNCA.)

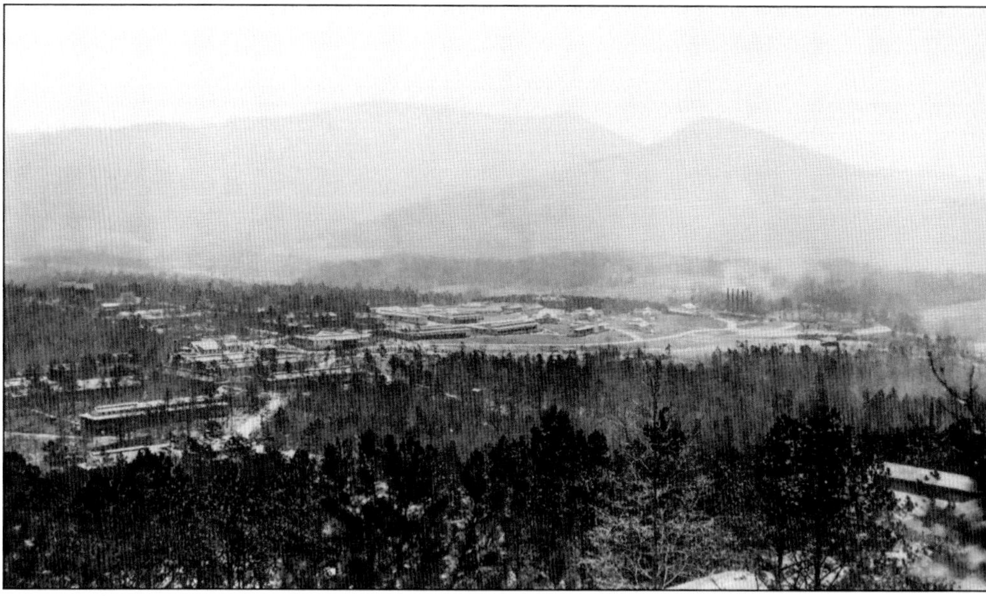

On March 25, 1918, ground was broken for the US Army General Hospital No. 19. The first patient was admitted on September 4, 1918. The Oteen Fire Department probably started at this time. In April 1919, construction of 104 buildings on 337 acres was completed. On October 16, 1920, the Army transferred the hospital to the US Public Health Service. (Courtesy of Charles George VA Medical Center.)

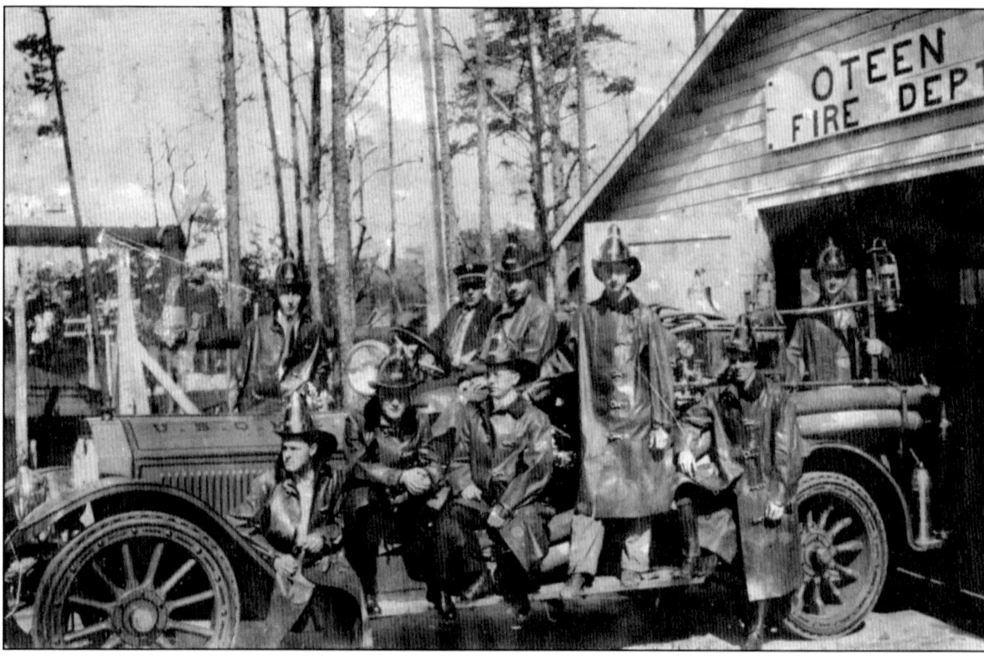

The site had 1,200 patients, 800 employees, a post office, and a railroad station named Azalea. About the time of the transfer, Colonel Hoagland gave it the name Oteen, a Cherokee saying meaning "Chief Aim" (which was to get well). On May 1, 1922, the hospital was transferred to the newly formed Veterans Administration with 1,000 beds primarily for the treatment of tuberculosis, 1,650 staff, 42 buildings, and 10 Quonset huts. (Courtesy of Bob Bryson.)

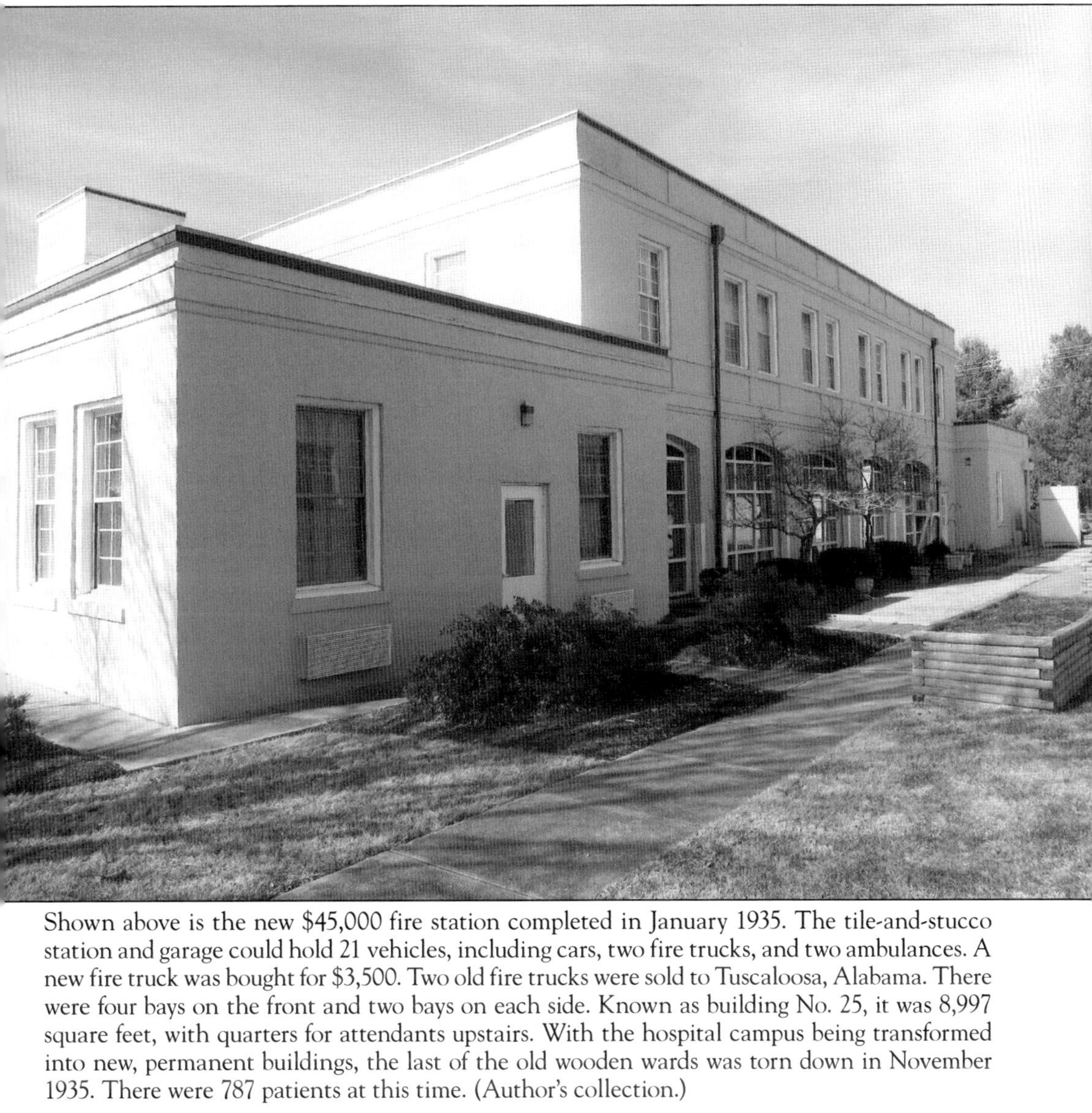

Shown above is the new $45,000 fire station completed in January 1935. The tile-and-stucco station and garage could hold 21 vehicles, including cars, two fire trucks, and two ambulances. A new fire truck was bought for $3,500. Two old fire trucks were sold to Tuscaloosa, Alabama. There were four bays on the front and two bays on each side. Known as building No. 25, it was 8,997 square feet, with quarters for attendants upstairs. With the hospital campus being transformed into new, permanent buildings, the last of the old wooden wards was torn down in November 1935. There were 787 patients at this time. (Author's collection.)

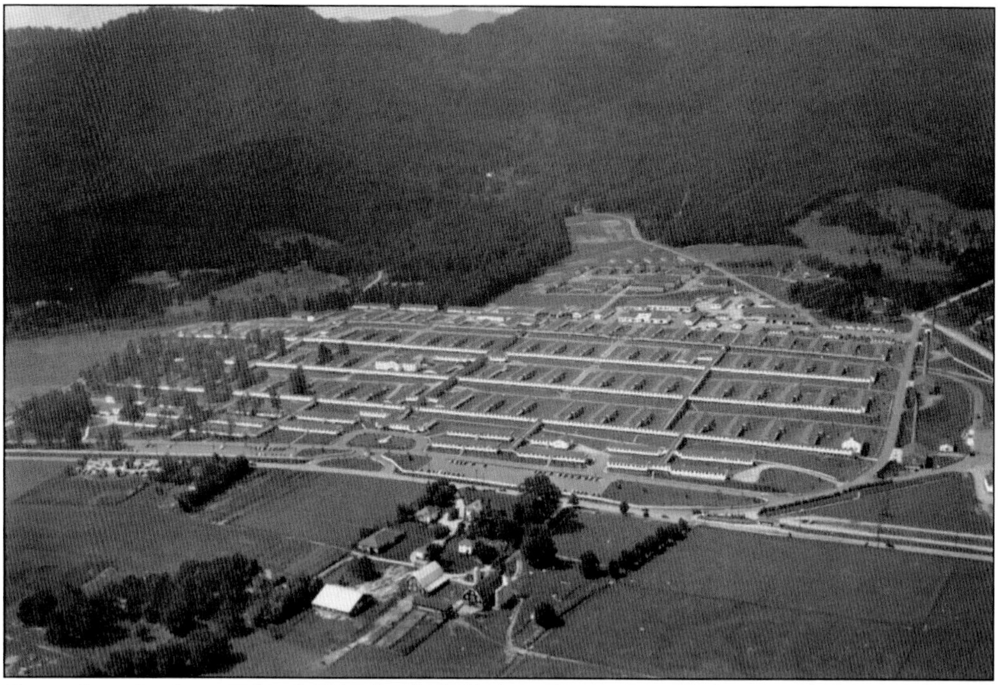

Ground was broken in April 1942 for the US Army Moore General Hospital, first named the Swannanoa General Hospital, on the grounds of the old Mountain Experiment Station of the State Test Farm. The first patient was admitted on December 2, 1942. The campus grew to 197 buildings on 264 acres, and a two-bay fire station was built at the rear center. The buildings were protected by sprinkler systems and a fire-alarm system that was connected to the fire station using an American District Telegraph alarm system. On November 15, 1946, the Veterans Administration took over operation of Moore General and 1,600 patients. Moore had been a tuberculosis center for several months before the transfer. Before that, it was a tropical disease center. Oteen and Swannanoa were under one management team. (Both, courtesy of Charles George VA Medical Center.)

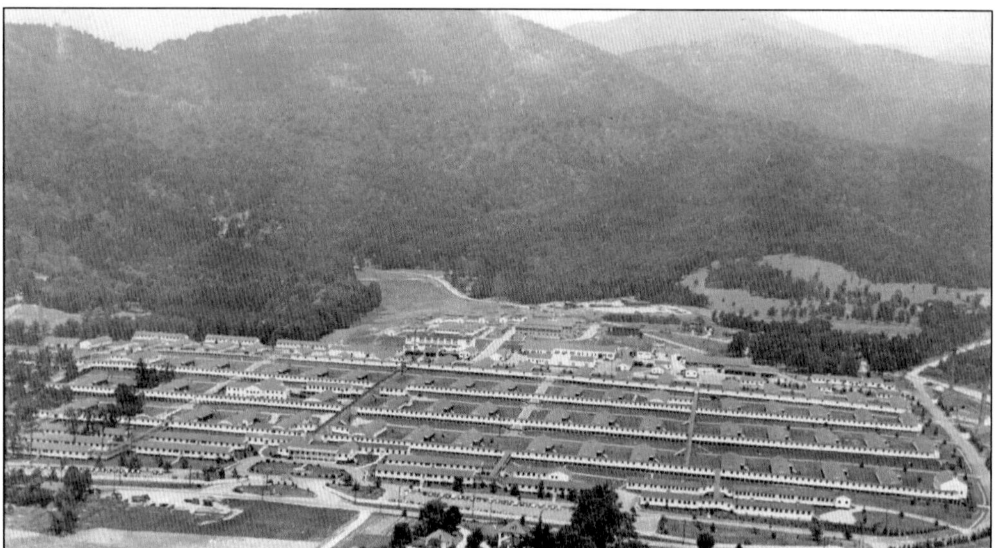

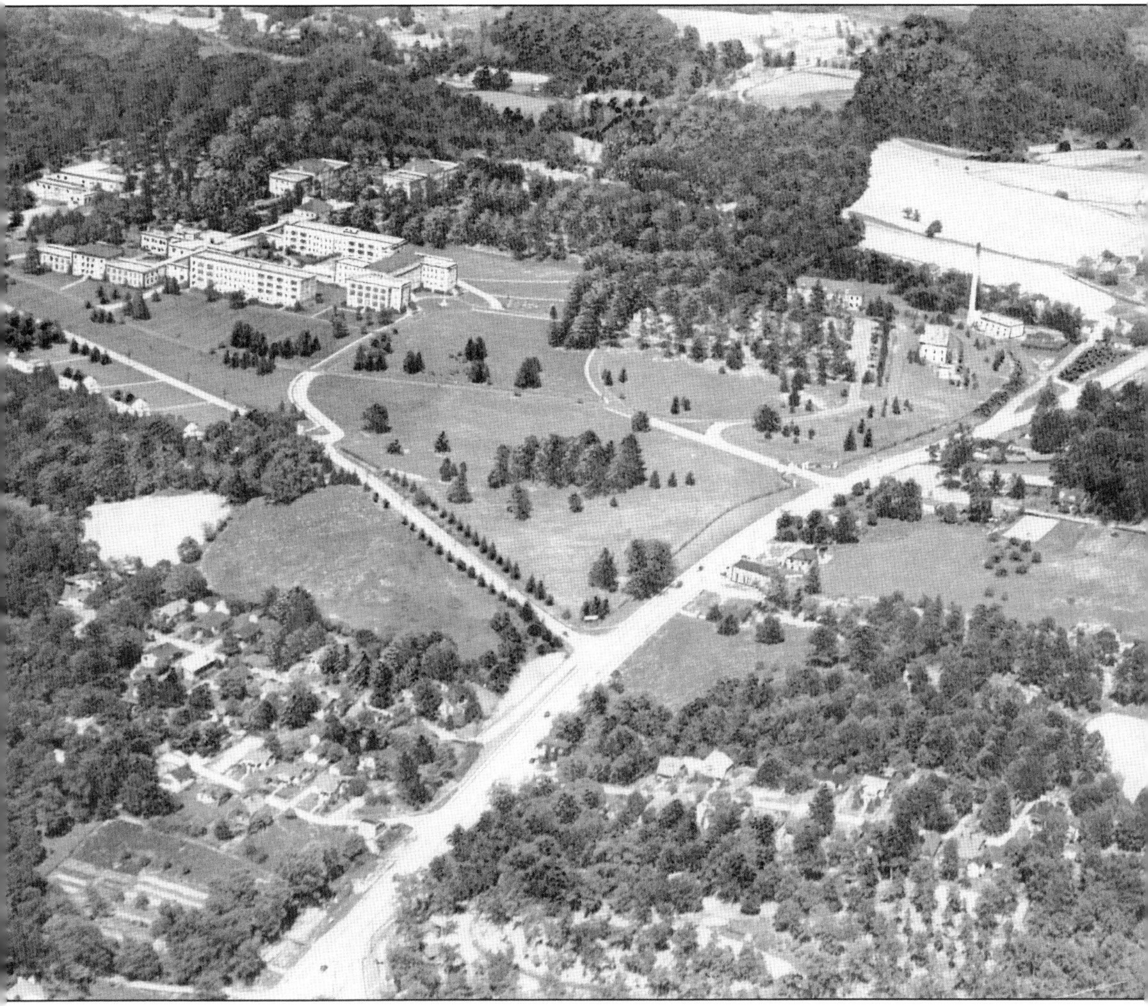

Bed patients were kept at Oteen's 300-acre campus, then transferred to Swannanoa when they became ambulatory. In September 1960, Moore General was turned over to the state of North Carolina to become the Juvenile Evaluation Center. The Juvenile Evaluation Center did have a fire department in the early 1960s. In June 1967, a new hospital was completed. It was built directly in front of the old hospital building seen in the center of the above picture. The old hospital buildings were sold except for one 125-bed unit. The department moved into their third station at this time, which was originally a spray-paint shop next to the laundry (small building on the right in front of the smokestack). (Courtesy of Charles George VA Medical Center.)

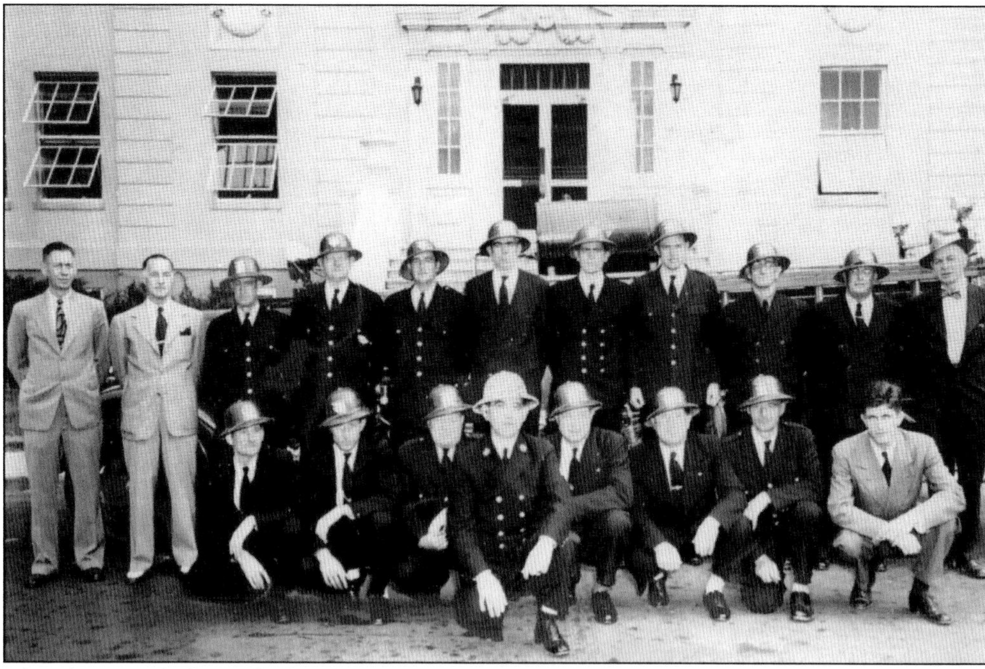

This photograph shows the fire department around 1935 with their truck in front of the old main hospital. At the time of the move, the fire department had 12 firefighters and one fire truck. They were primarily security guards, working 24 hour shifts, and it is said that they were really good cooks. In 1976, the fire department closed, and the VA contracted with the City of Asheville for fire protection. (Courtesy of the Charles George VA Medical Center.)

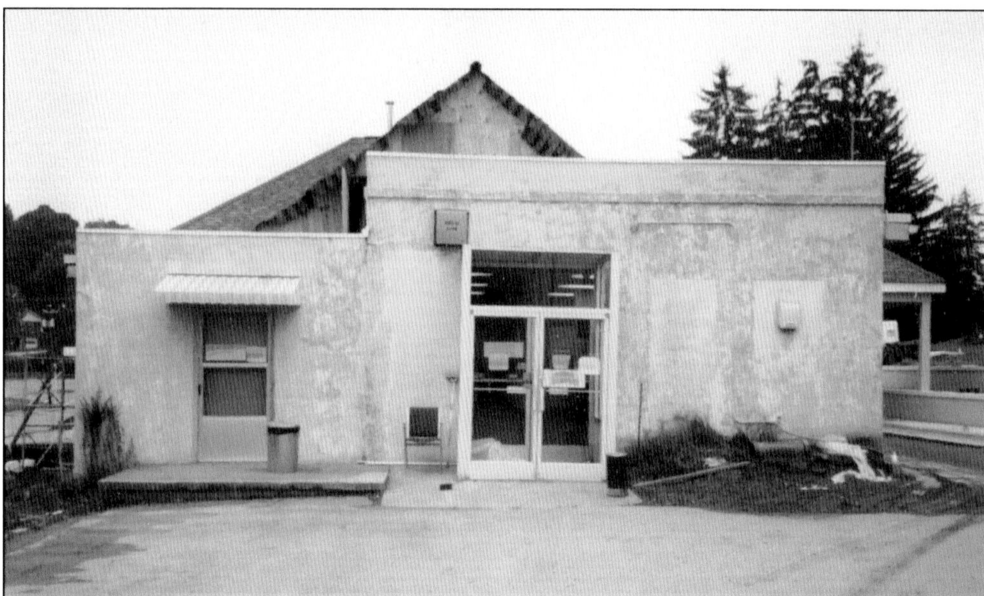

The fire station became the credit union when they started remodeling for that use in May 1977. The image above shows remnants of where old windows (on the right) and the apparatus bay (double doors in the middle) used to be before the credit union built an addition and remodeled the exterior. (Photograph by Shelia Biddix, courtesy of Charles George VA Medical Center.)

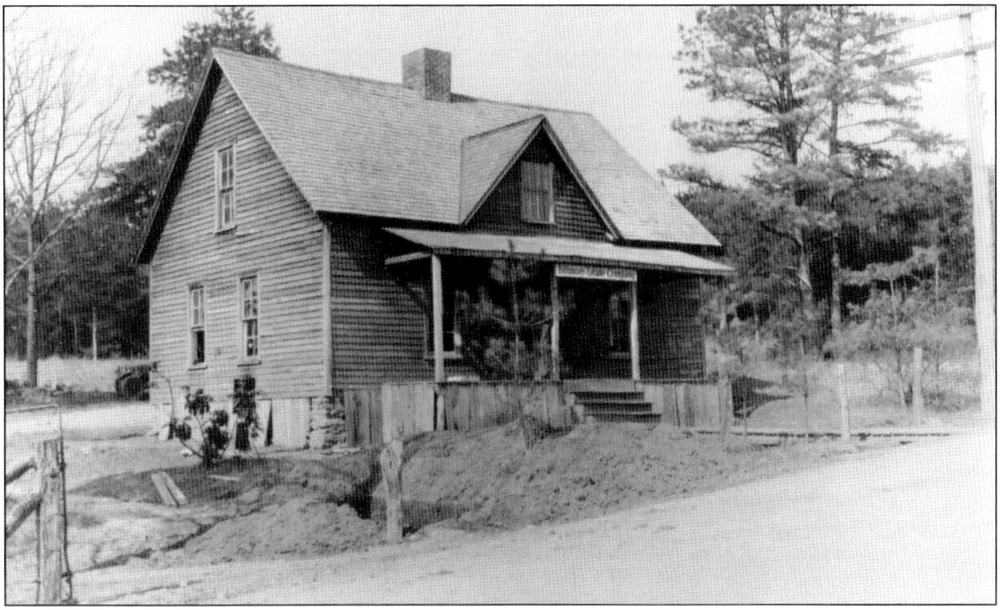

August Junius Adams was requested by Edith Vanderbilt to study ways to reduce maintenance costs after the flood of 1916 damaged the Biltmore Village and Estate. Adams recommended developing 1,500 acres of unproductive and high-tax land along Hendersonville Road. The plan was approved, but it was decided not to proceed until after World War I. This house at Busbee and Hendersonville Roads was the first real estate office. (Courtesy of the Town of Biltmore Forest.)

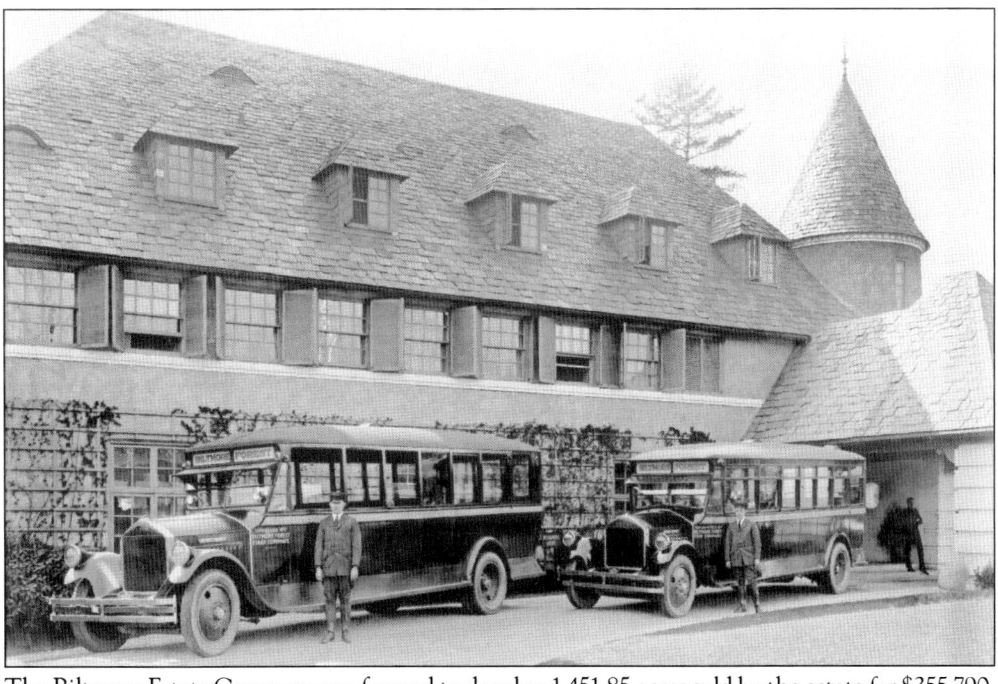

The Biltmore Estate Company was formed to develop 1,451.85 acres sold by the estate for $355,790, with $20,000 in stock and $335,000 in first-mortgage, 15-year, sinking-funds bonds. The estate lent $100,000 to build the above country club and golf course. As pictured above, Biltmore Forest did have its own bus service for a time. (Courtesy of the Town of Biltmore Forest.)

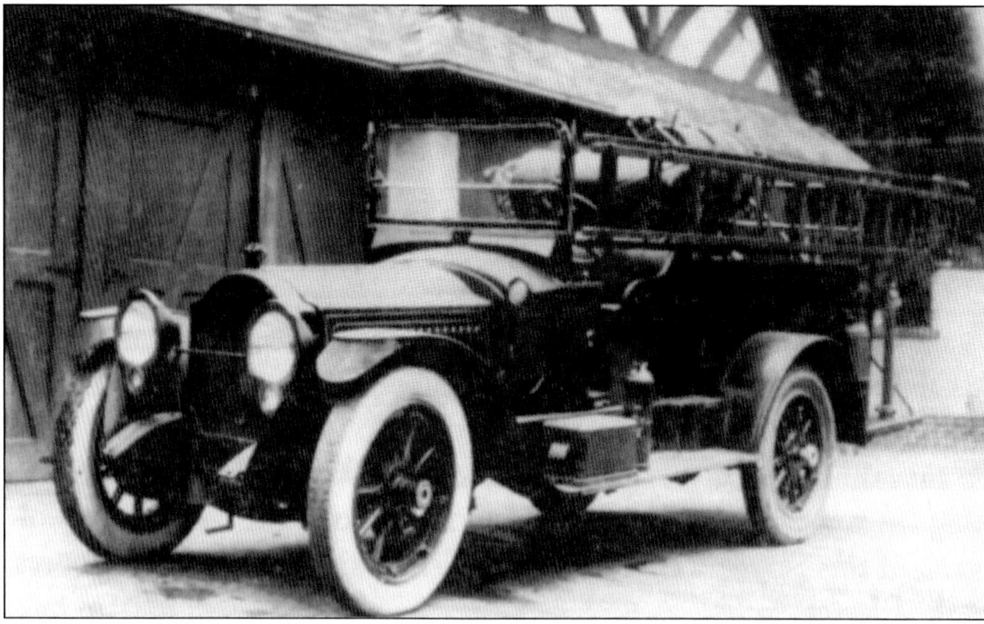

In 1921, the Public Works building was constructed; it housed the first fire truck, shown above mounted on a Packard chassis. In 1923, the town of Biltmore Forest was officially chartered. The first meeting was held at the real estate office on March 20, 1923, and the first election was held on the first Monday in May. At the meeting, a committee was formed for the fire department. (Courtesy of the Town of Biltmore Forest.)

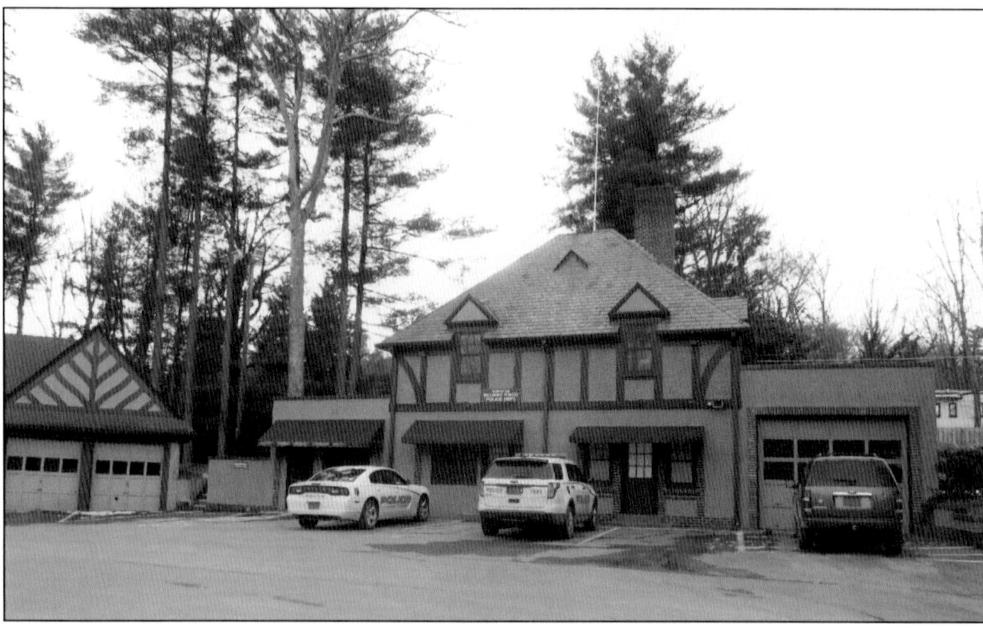

In 1923, Charles Parker designed the first station with an upstairs apartment for the chief of police. C.W. Hensley was the first chief and was paid $90 a month for his services. The original apparatus bay was in front of the Dodge police car. The left wing was possibly added on in 1935 through the Federal Emergency Administration of Public Works. The right wing was added in 1977. (Author's collection.)

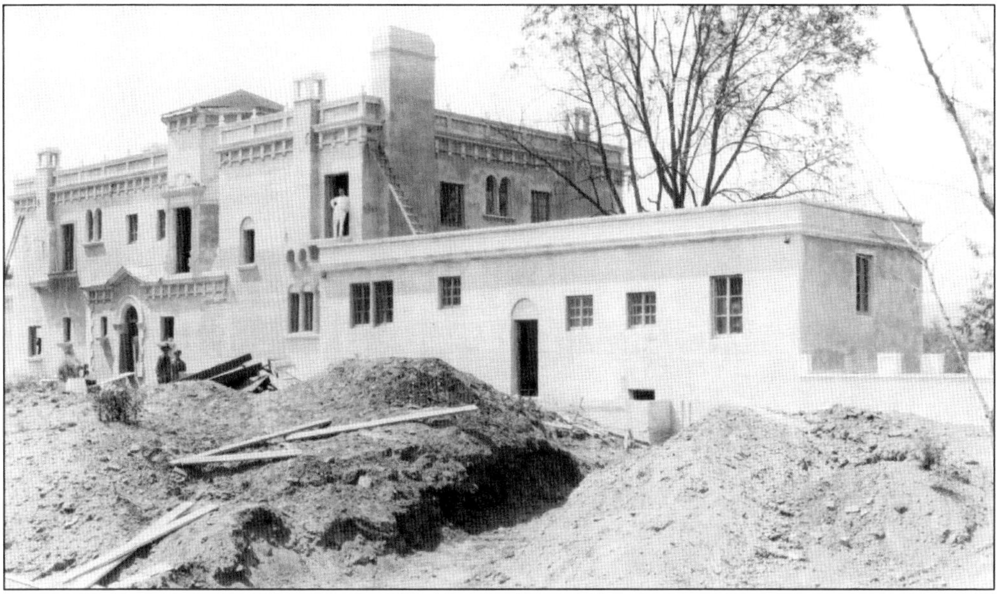

A new American LaFrance Type 91, 500-gallon-per-minute, 80-gallon chemical tank arrived in February 1929. On July 28, 1959, a new $14,000 Ford American LaFrance fire truck with a 750-gallon-per-minute pump and 500-gallon tank arrived complete with hose, aluminum ladders, and two Scott Airpacks. Six policemen and eight volunteers were on the department at the time. Shown above is the Frith, Edith Vanderbilt's home in Biltmore Forest. (Courtesy of the Town of Biltmore Forest.)

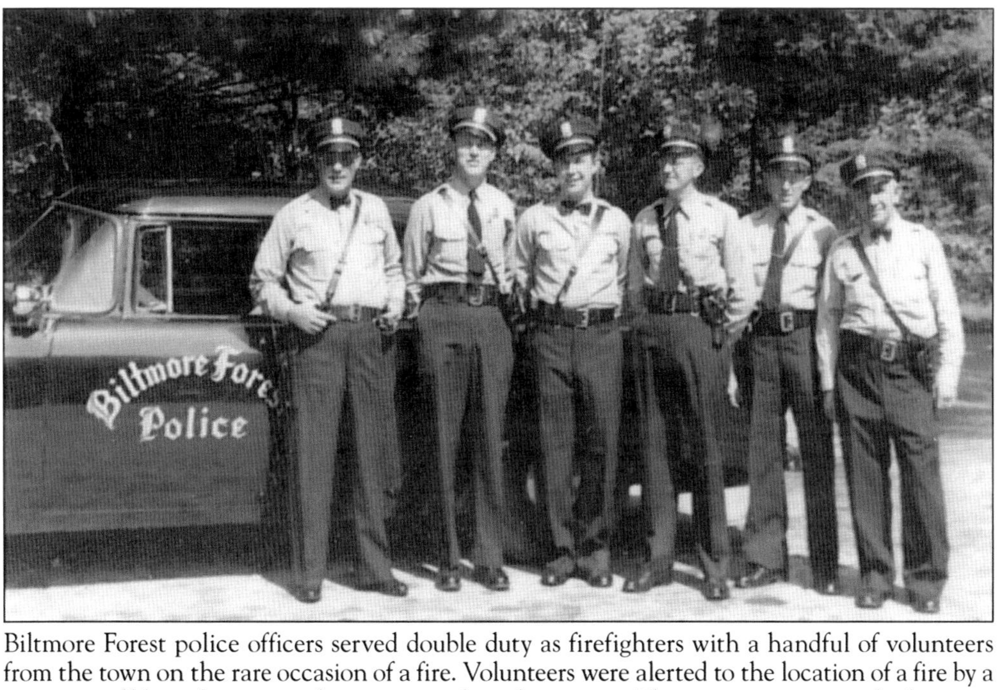

Biltmore Forest police officers served double duty as firefighters with a handful of volunteers from the town on the rare occasion of a fire. Volunteers were alerted to the location of a fire by a sequence of blasts from an air horn mounted on the station. The station was a popular hangout for town children and adults. In 1981, duties were first split and a new Ford Grumman truck arrived. (Courtesy of the Town of Biltmore Forest.)

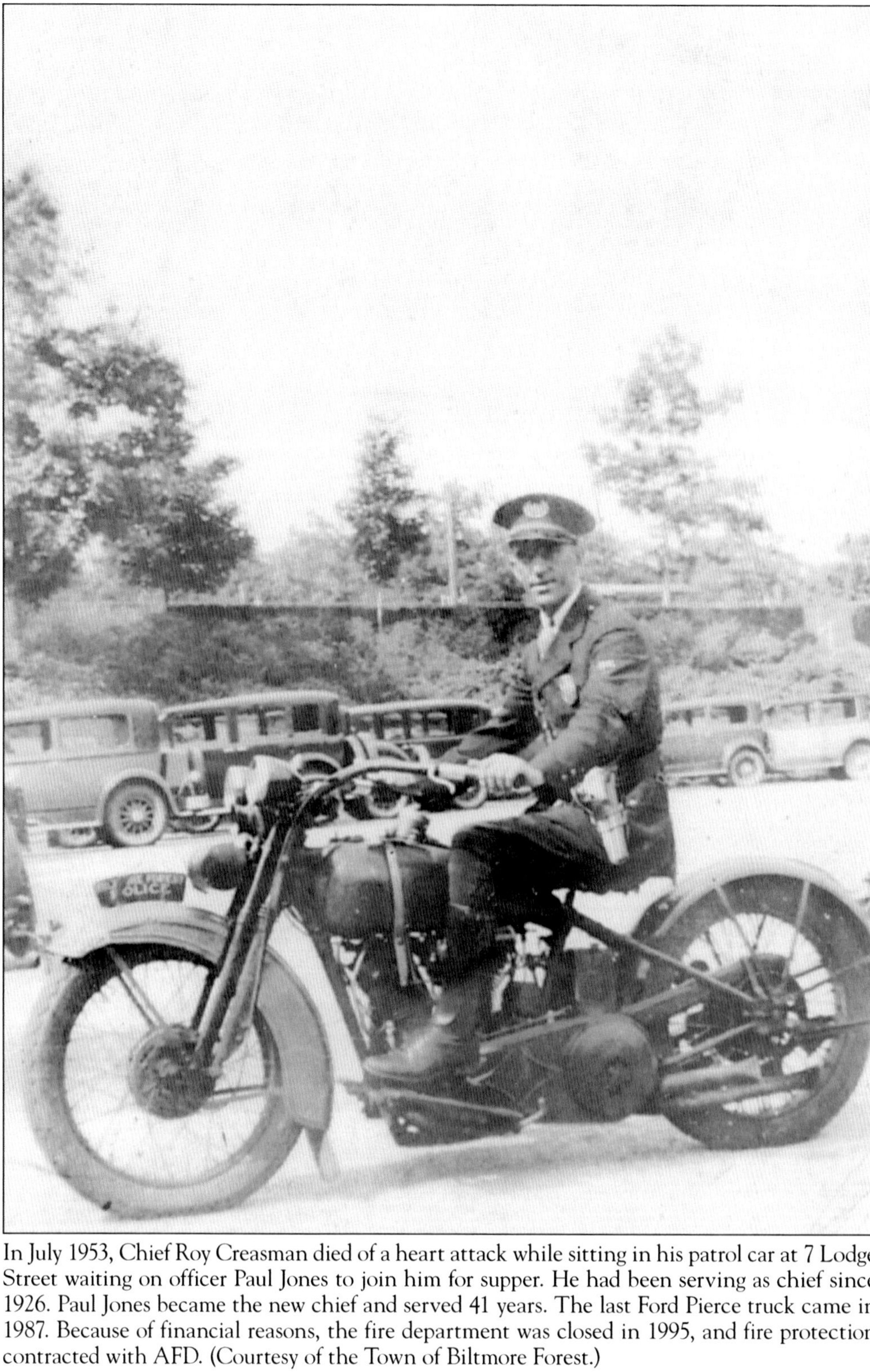

In July 1953, Chief Roy Creasman died of a heart attack while sitting in his patrol car at 7 Lodge Street waiting on officer Paul Jones to join him for supper. He had been serving as chief since 1926. Paul Jones became the new chief and served 41 years. The last Ford Pierce truck came in 1987. Because of financial reasons, the fire department was closed in 1995, and fire protection contracted with AFD. (Courtesy of the Town of Biltmore Forest.)

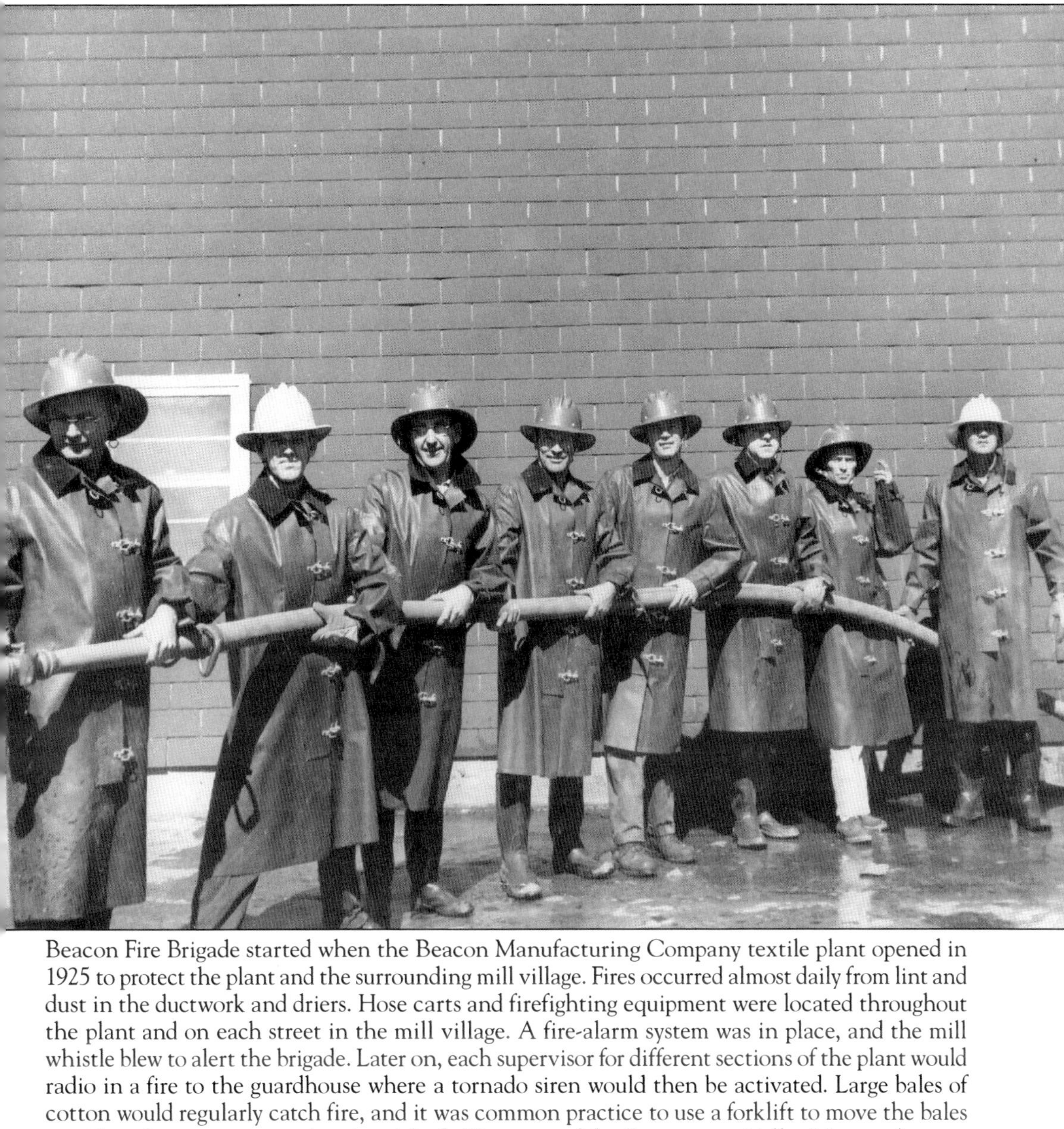

Beacon Fire Brigade started when the Beacon Manufacturing Company textile plant opened in 1925 to protect the plant and the surrounding mill village. Fires occurred almost daily from lint and dust in the ductwork and driers. Hose carts and firefighting equipment were located throughout the plant and on each street in the mill village. A fire-alarm system was in place, and the mill whistle blew to alert the brigade. Later on, each supervisor for different sections of the plant would radio in a fire to the guardhouse where a tornado siren would then be activated. Large bales of cotton would regularly catch fire, and it was common practice to use a forklift to move the bales outside to be torn apart and extinguished. (Courtesy of the Swannanoa Valley Museum.)

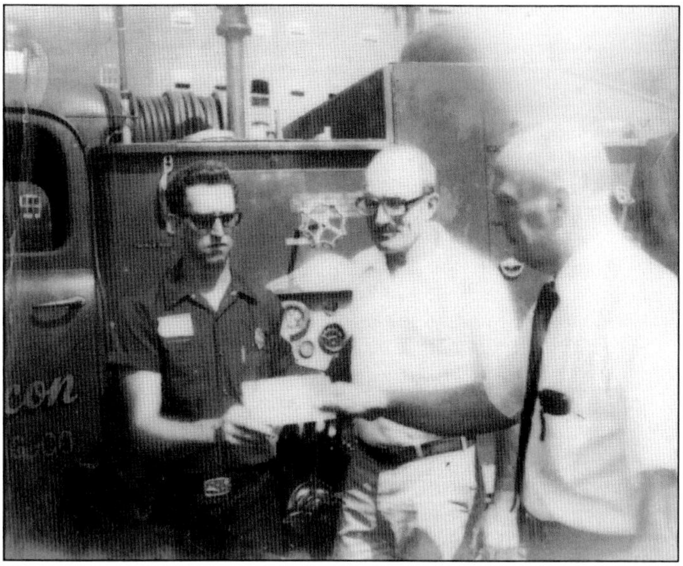

Swannanoa eventually grew too big for the fire brigade to handle, so the plant assisted in the formation of the Swannanoa Fire Department in 1959. Here, Beacon is presenting its one and only fire truck, a Dodge equipped with foam spray and a dry chemical unit, to members of the Swannanoa Fire Department sometime around 1976. The mill can be seen in the background. (Courtesy of Swannanoa Fire Department.)

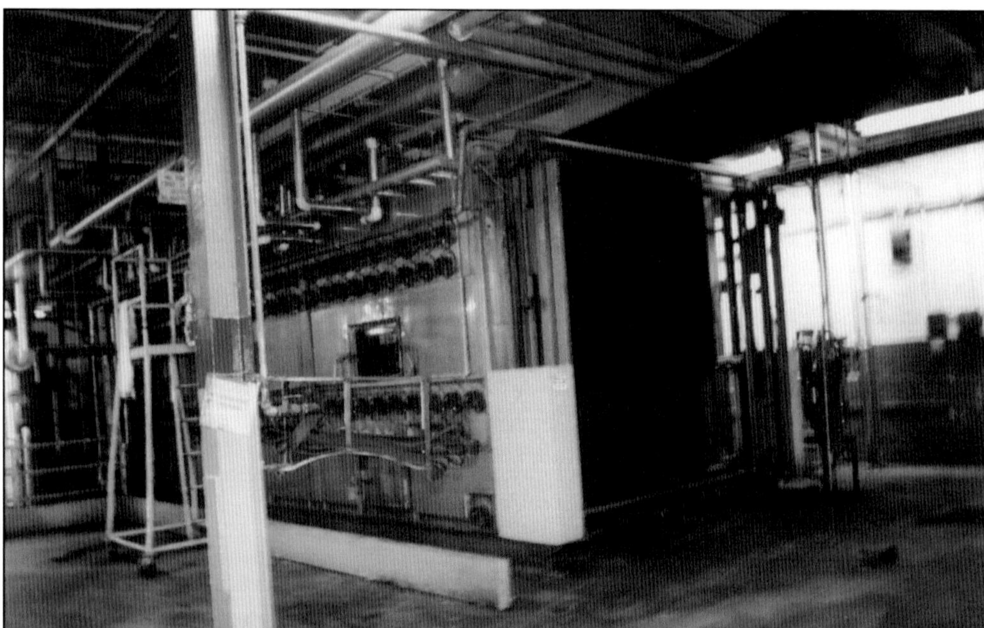

Beacon had a fire marshal that would walk the entire plant daily looking for fire hazards, maintaining the sprinkler system and 21 fire pumps, issuing welding permits, and alerting the fire brigade when welding was happening. The most interesting task was checking the water-storage ponds under the plant. A boat was used to row over to the intakes that cooled machines and supplied the sprinkler system. (Courtesy of the Swannanoa Valley Museum.)

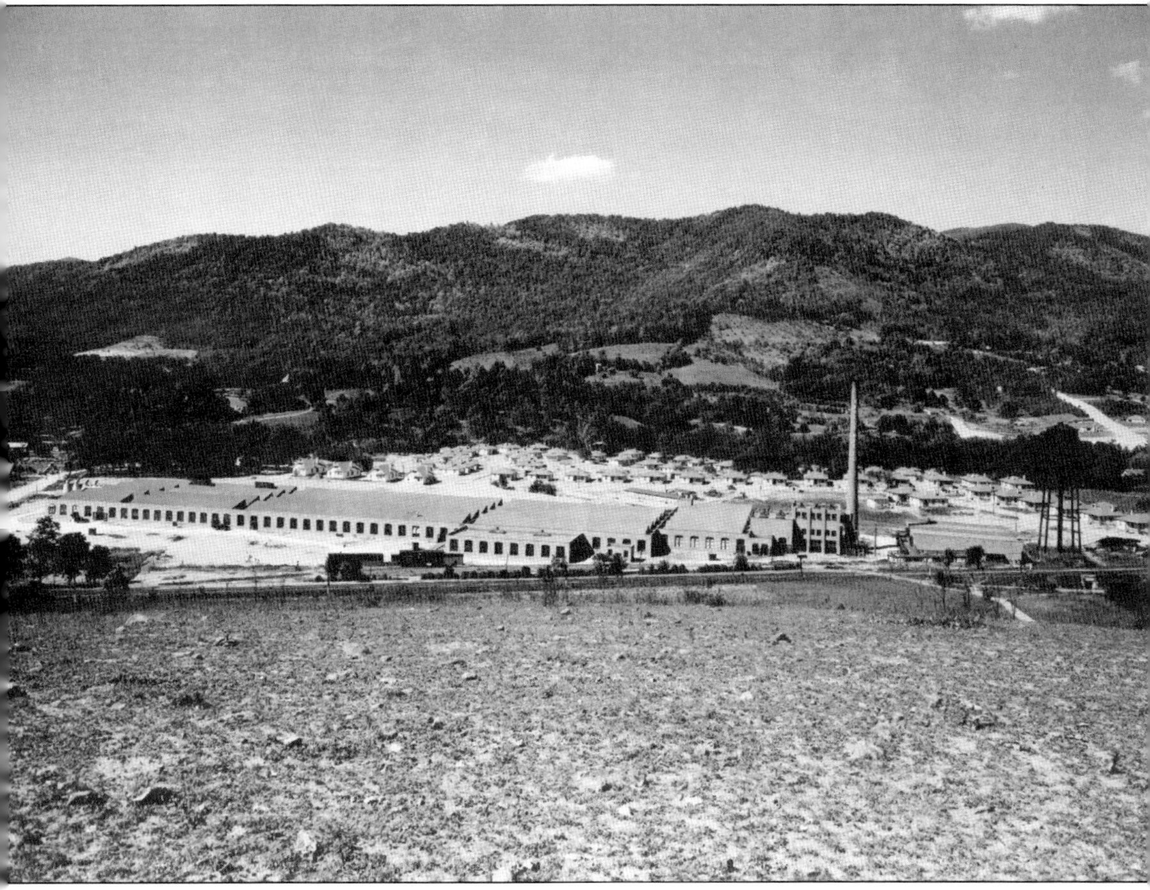

Beacon Manufacturing started in 1904 in New Bedford, Massachusetts, making blankets. Charles D. Owen II was looking to expand his company into the south, and he spotted the 160-acre Whitson farm from the train he was traveling on. Whitson sold his farm to Owen with the agreement that his son would have a job at the plant. That son would go on to retire from the plant. Beacon was the first business in the Swannanoa Valley and would grow to include 1,600 acres; the mill covered 42. Starting in 1924, Beacon Village was built for employees and included three separate phases of housing, a company store for groceries and clothing, a gas station, and a bank. During the Great Depression, 85 percent of the textile mills in New Bedford closed. In 1933, to escape rising labor costs and forming unions, Owen moved the entire New Bedford mill—all the machines and even the bricks—by train in what would be the first and largest move of the textile industry. (Courtesy of North Carolina Collection, Pack Memorial Public Library, Asheville.)

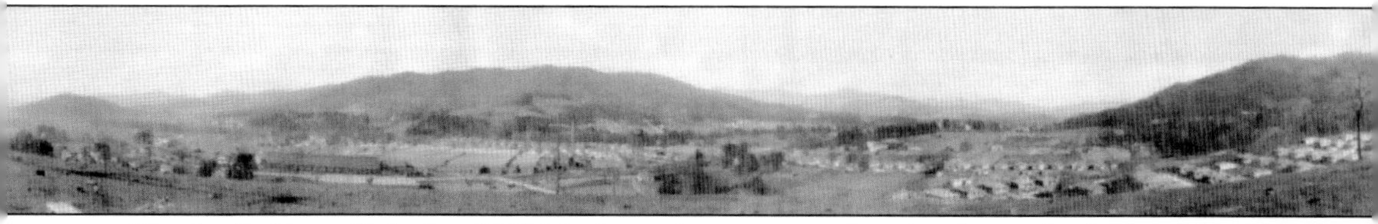

Charles D. Owen II died in 1937. His funeral train stopped at the plant, which closed for a brief moment to allow everyone to walk through and pay their respects. Charles D. Owen III carried on the company. During World War II, Beacon converted to all wool blankets for the military and contributed the largest number of its workforce (930 of 2,200 employees) to the war, more than any other industrial company in the nation. Over the years, it had a baseball, basketball, and golf team playing in the textile industrial league. The company would grow to include seven mills, and Beacon was at one time being the largest weaving room in the world, with 1,200 looms. In 1964, the plant was sold to National Distillers. It changed hands a few times and downsized before closing on April 15, 2002, with 300 workers. Five generations of family worked at the plant. The fire brigade ended at this time. (Courtesy of Swannanoa Valley Museum.)

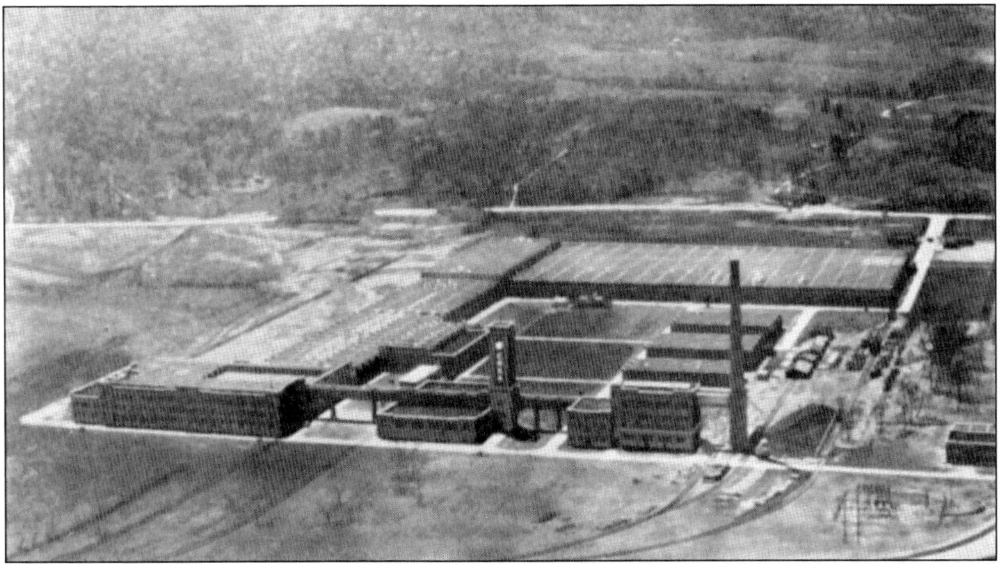

Ground was broken on October 1, 1928, for the American Enka Corporation plant. One mile of Hominy Creek was moved, the site was raised four feet, and 80 acres were dammed to create the 365-million-gallon Enka Lake. It opened on July 1, 1929, making rayon yarn and employing 2,200 people. The plant covered 75 acres, with 25 of those acres under roof. The company also owned another 2,000 acres outside the plant. It had fire and police departments, a hospital, and 100 homes for employees. The ENKA Fire Department started at this time with the below Ford truck, which had a 30-gallon chemical tank. Joseph R. Gill was the first chief. The department covered five districts, with the plant's steam whistle blown to indicate which the call was in. The steam whistle blew twice to indicate the fire was out. At night, telephones were used instead. (Both, courtesy of Enka-Candler Branch Library.)

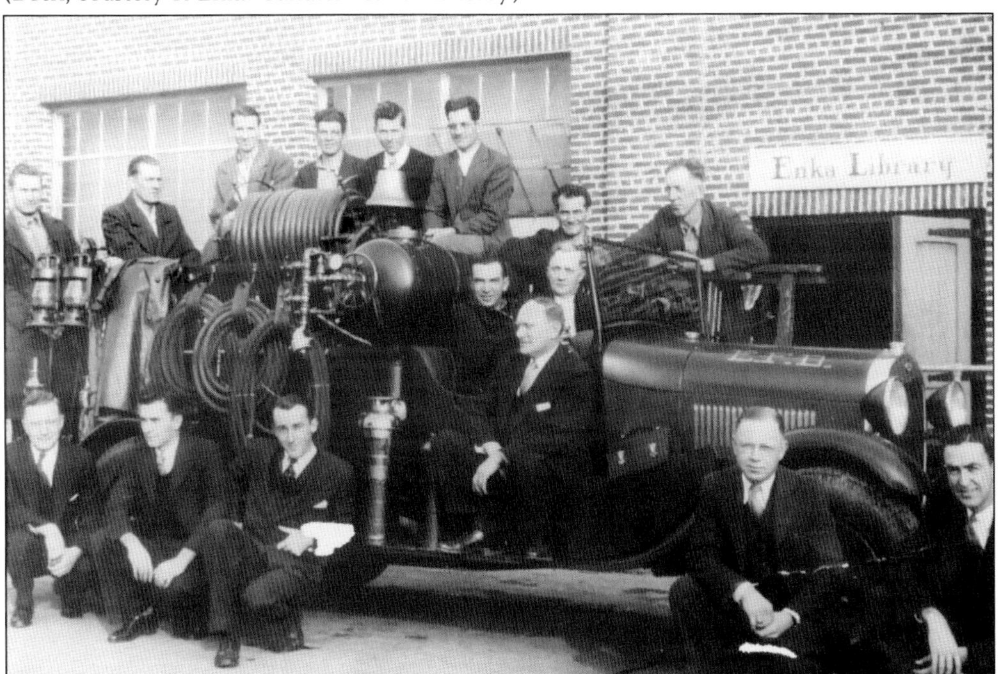

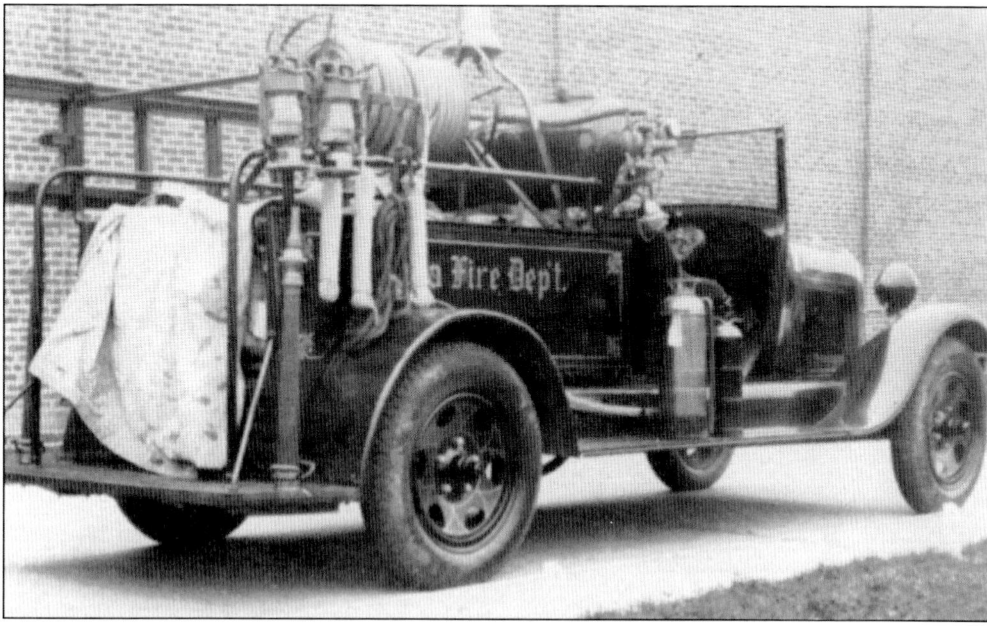

There were five companies with 7 to 21 men in each. Nearly all of them lived in the plant village. Each company was assigned a district. The chemical district company (Company C) knew about water reactions and responded only to that area unless off duty. Police Company D responded with the fire truck and assisted as directed by the chief. Members of Power Company E stayed at the powerhouse and operated the fire pumps except for one man that responded to the scene and was responsible for water supply. The shift electrician and chemist responded as well, with the chemist bringing gas masks and canisters. E.M. Salley was the assistant chief and would go on to serve as chief from 1945 to 1960. In June 1934, a Foamite Foam Generator, shown below, was added with a claim that it was the only one in Western North Carolina. (Both, courtesy of Enka-Candler Branch Library.)

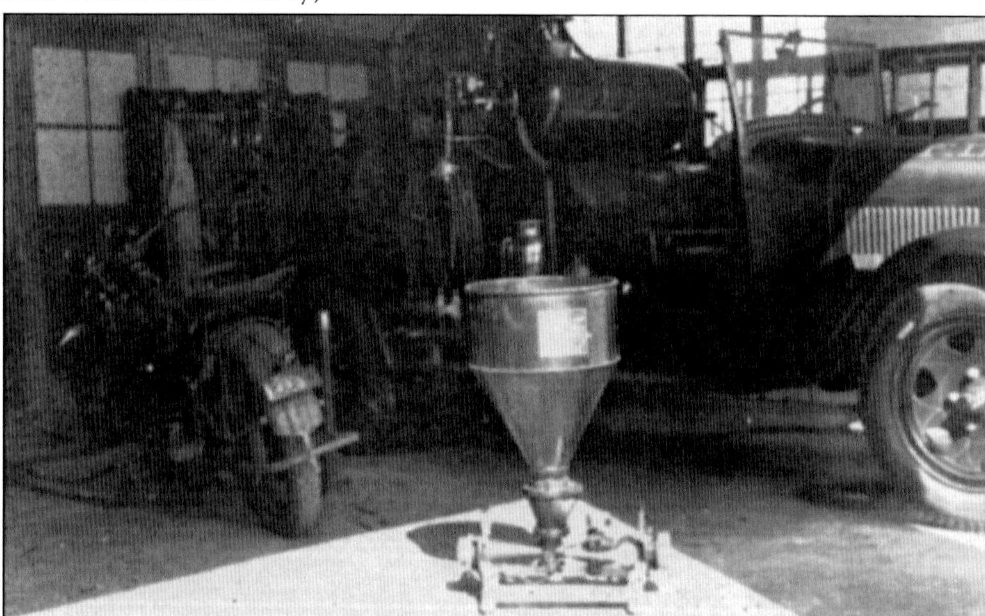

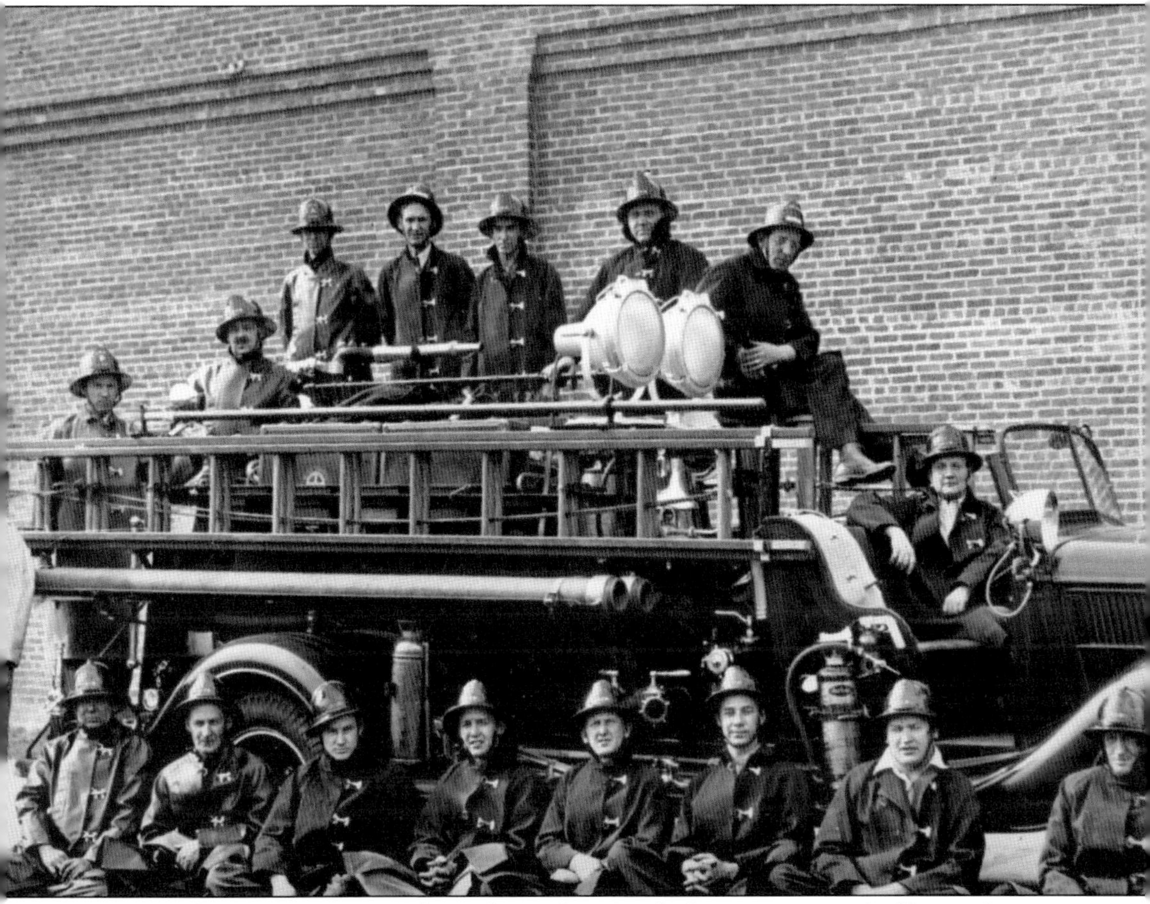

A 100,000-gallon water-storage tank was located in the clock tower to provide 60 pounds per square inch to the sprinkler system. At the plant were a 3,600-gallon-per-minute, motor-driven fire pump and a 3,600-gallon-per-minute, steam-driven fire pump. Water sources included Enka Lake, the 33,000-gallon-per-minute Hominy Creek, and City of Asheville water available in the plant and village. Extinguishers and ladders were located on garages around the village. (Courtesy of Enka-Candler Fire Department.)

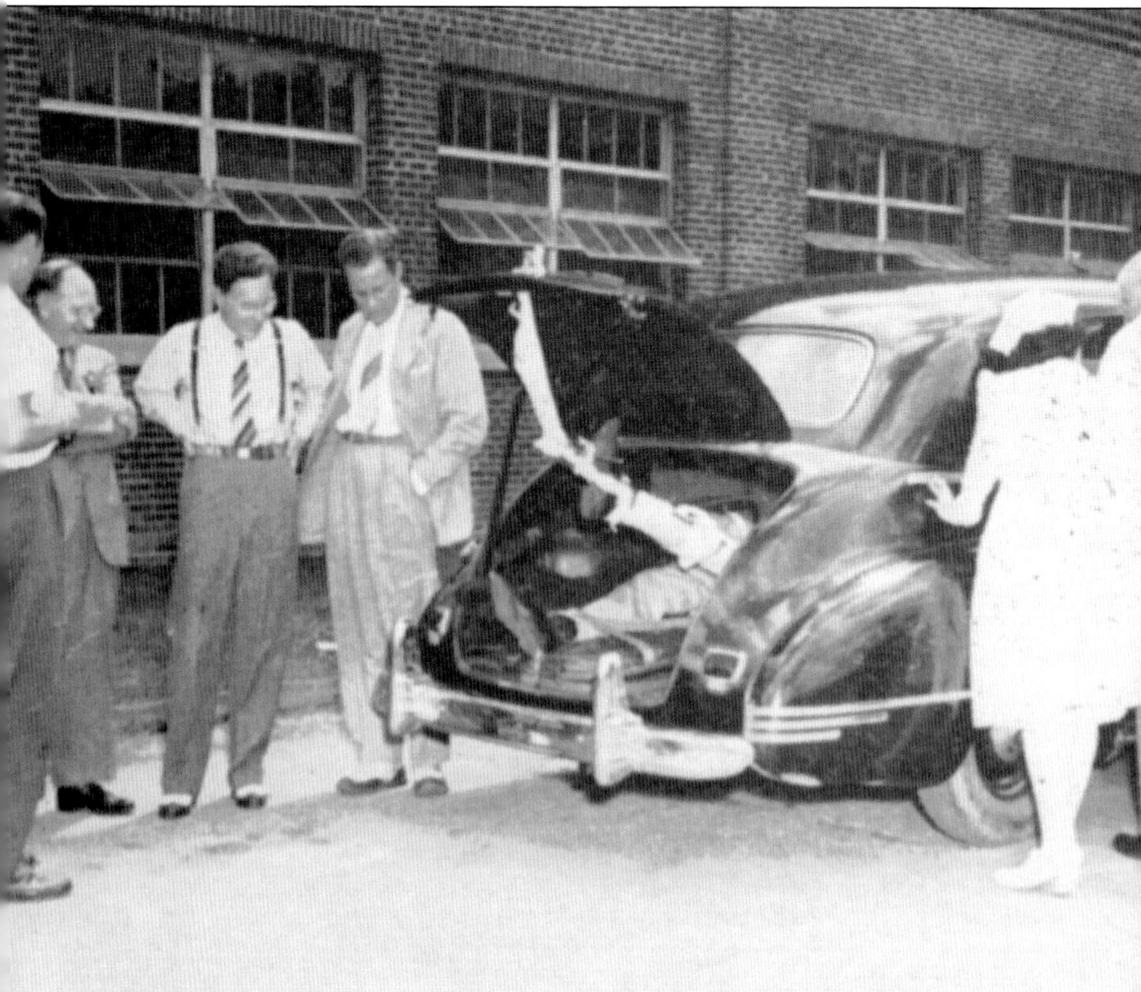

In September 1945, the above stretcher service was available. The plant provided a library, a dental clinic, a general store, a recreation club, a church, a drugstore, a barbershop, a post office, a beauty parlor, and sports teams. After 20 years, there were 22 firefighters in two companies. The Hose Company fought the fire, and the Salvage Company raised ladders, covered furniture, and cleaned up. In 1951, they had 9 alarms in the plant and 40 outside. Two major floods happened, one in 1940 and the other in 1977. (Courtesy of Enka-Candler Branch Library.)

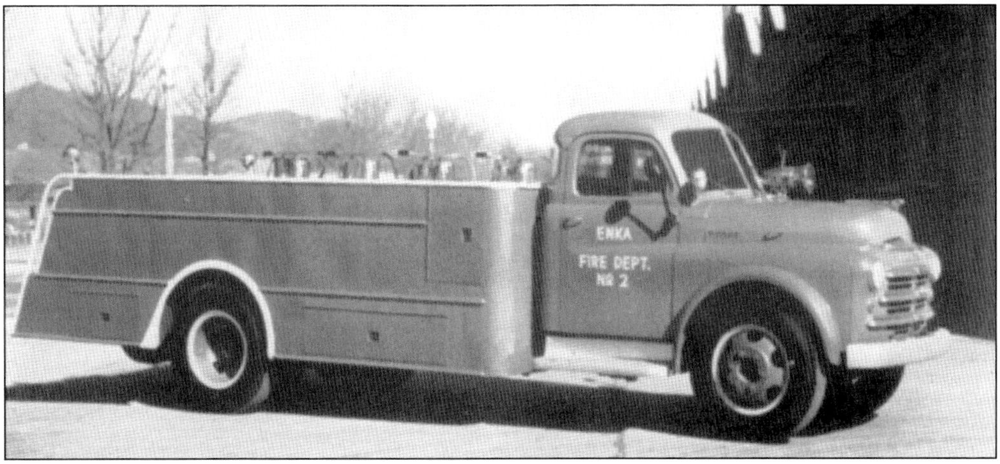

In March 1951, the above 1,000-gallon truck with 700 feet of hose was designed by the engineering department and built by the maintenance department for rural firefighting and street flushing. In 1951, ENKA had 50 employees at their 206 Madison Avenue office in New York. The office was in a 26-story building and was home to the company president as well as the sales, engineering, and research and analysis departments. They served 300 customers. (Courtesy of the Enka-Candler Branch Library.)

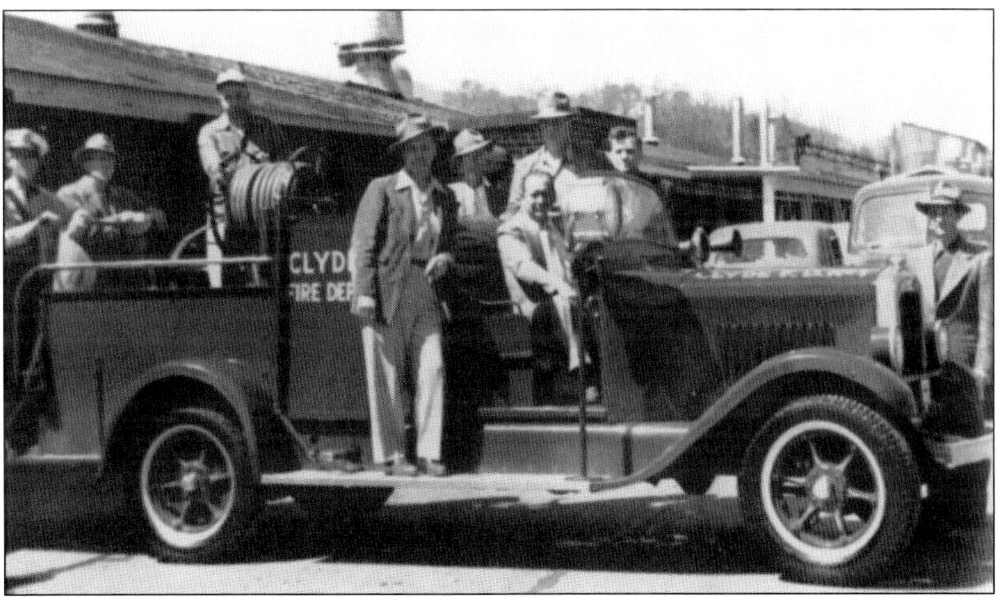

The above truck was ready in May 1953 for Clyde to start a fire department. Plant painters and mechanics in the auto-repair shop spent $300 on refurbishing a 1925 GMC fire truck from Weaverville. A large number of plant employees lived in Clyde. The Asheville Fire Department covered calls for ENKA when they were answering other calls. Firemen drilled each week. (Courtesy of the Enka-Candler Branch Library.)

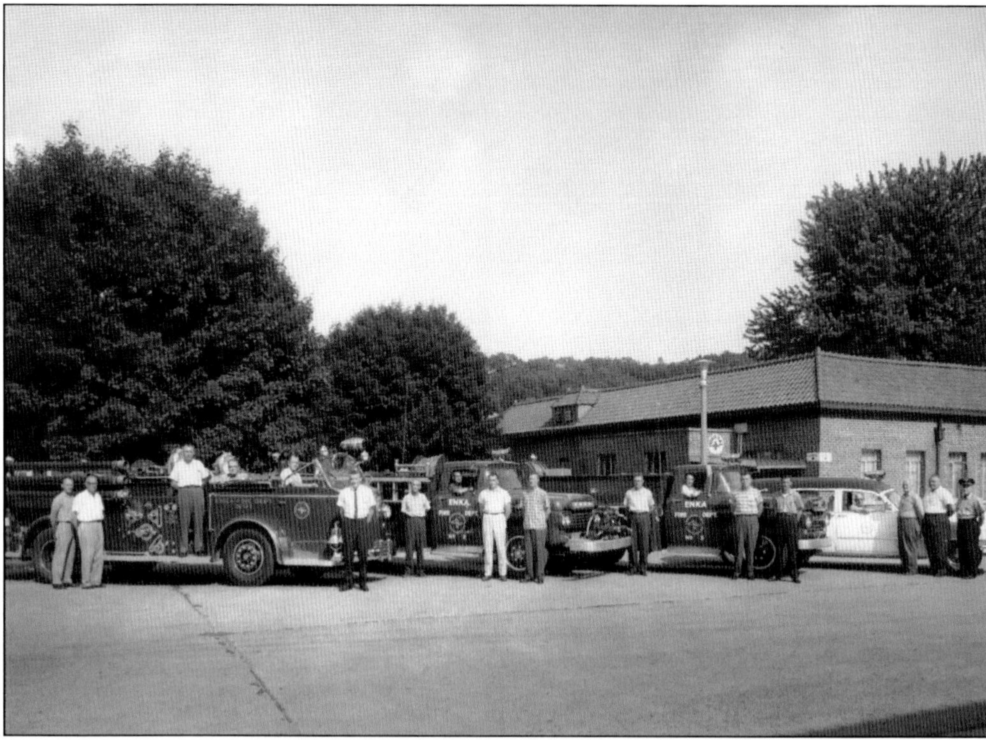

The plant grew over the years, starting with 2,200 employees, expanding to 2,850 a decade later, and reaching 4,164 by May 1947. Employees were brought in by a 2-8-2 steam train from Canton, by bus or car, or by walking from the nearby village. A four-mile fire district was created in 1958, contracted by Buncombe County for fire protection. The fire department at the time had the three fire trucks pictured above and 24 firefighters. (Photograph by Allan Culberson; courtesy of Enka-Candler Fire Department.)

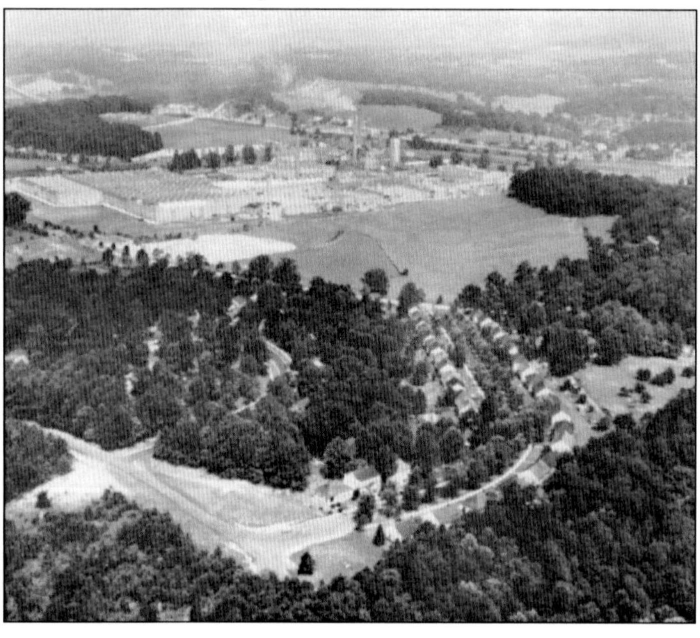

Shown are Enka Village and the plant. American Enka was purchased by BASF Fibers in 1985. In 1986, ENKA Fire Department answered 285 calls, with 85 percent being outside the plant. In December, BASF asked the Buncombe County commissioners to take over fire protection outside the plant because of decreased manpower at the plant and increased population. They would stop answering calls on December 31, 1987. (Courtesy of the Enka-Candler Branch Library.)

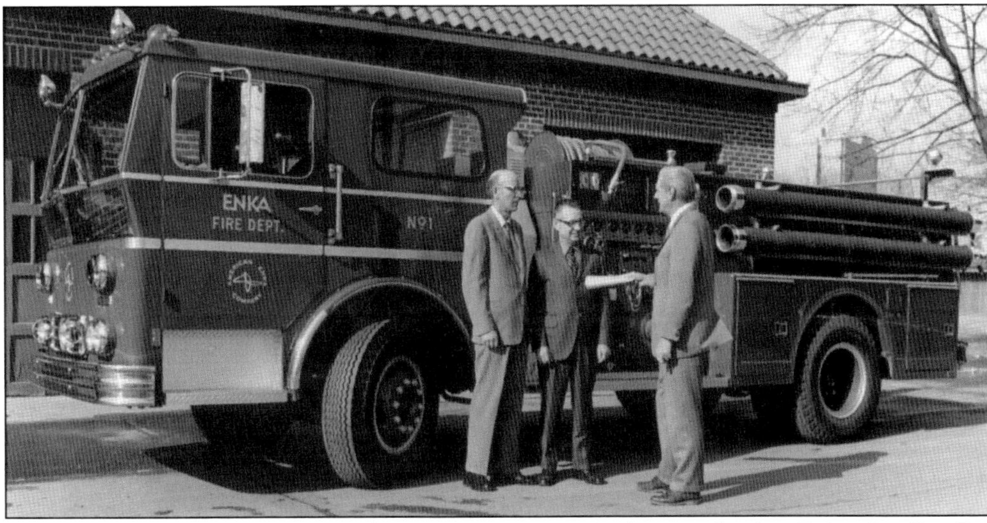

This Ward-LaFrance in front of the station was ENKA's last truck after Enka-Candler Fire Department formed on January 1, 1988. BASF's donation to Enka-Candler was valued at $250,000. Neil Davis was the last ENKA chief. He was part of the Shift Fire Response Team formed of shift personnel to extinguish fires in the beginning stages. The team remained for about three years, and the truck was then sent to ENKA's Lowland plant. (Courtesy of Enka-Candler Fire Department.)

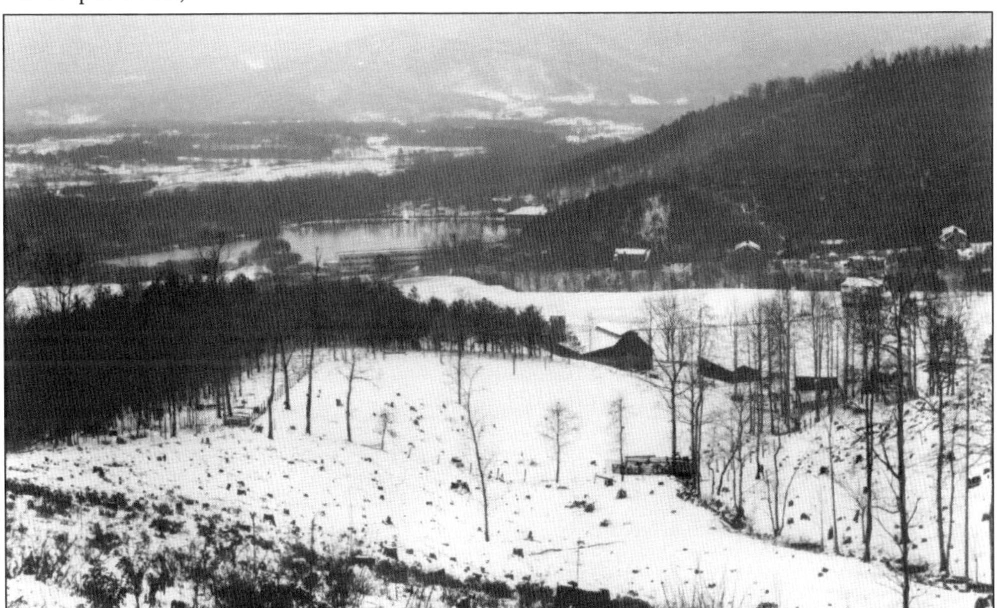

Black Mountain College was a progressive liberal arts school in operation from 1933 to 1957 that believed in art as a central focus for students. The creative mind of Buckminster Fuller was a faculty member. The students did operate a fire brigade. The campus was founded on an area first built by Edwin W. Grove as a recreation area for his planned community called Grovemont, complete with the man-made Lake Eden, dining and recreation hall, lodges, and cabins. (Photograph by Trude Guermonprez; courtesy of North Carolina Western Regional Archives Black Mountain College Research Project.)

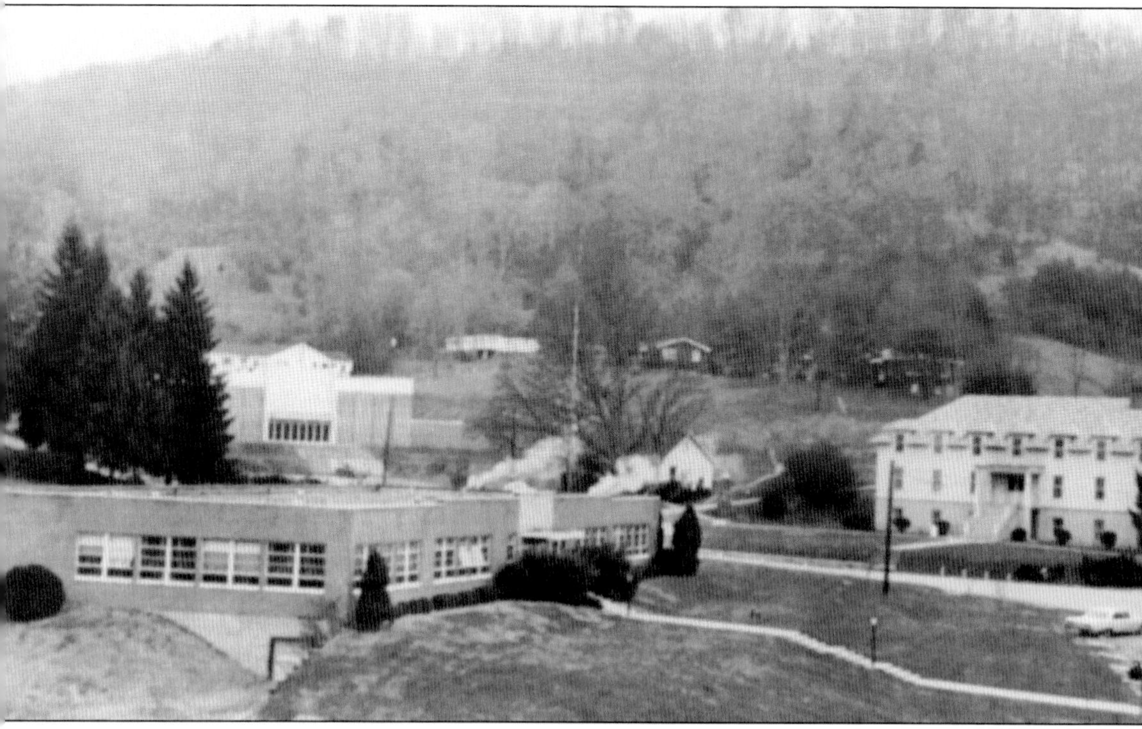

Mount Pisgah Academy started in June 1914 as the self-supporting Seventh-day Adventist Pisgah Industrial Institute boarding school. The campus would change over the years to include the above buildings. The three-story facility on the right was Fleetwood Hall (now demolished), built in 1936 from the disassembled materials of the failed 12-story Fleetwood Hotel on Jump Off Mountain near Hendersonville. The building in the left foreground was the 25-year-old administration building, which burned on November 4, 1983. The fire, which started in the library, was reported at about 4:30 a.m., and ENKA, Upper Hominy, and West Buncombe responded. Part of the roof had collapsed, and one end was heavily involved in flames when firefighters arrived. The building included offices, classrooms, a science lab, a library, and a computer room. Two firefighters were injured during the blaze. When a fire department existed here, the boys would answer calls that came in by running from their dorm (located above the gym on the left) across the field to the fire hall (located behind Fleetwood Hall), sometimes without clothes on to avoid having their bunker pants and boots slowing them down. (Courtesy of Mount Pisgah Academy.)

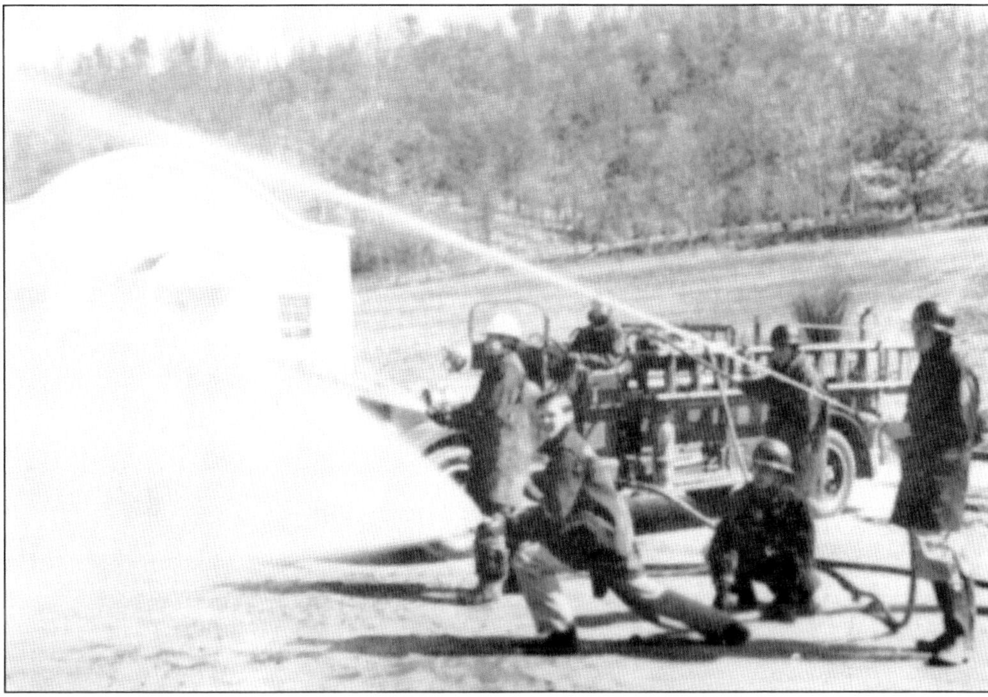

Faculty member Edward Reifsnyder thought it would be good to have a fire department, so he started the firemanship program in 1952. The program included about 14 students—very select juniors and seniors that could handle the extra workload, had good grades, were under no disciplinary actions, and were very mature—and faculty. This is their first fire truck. (Courtesy of Mount Pisgah Academy.)

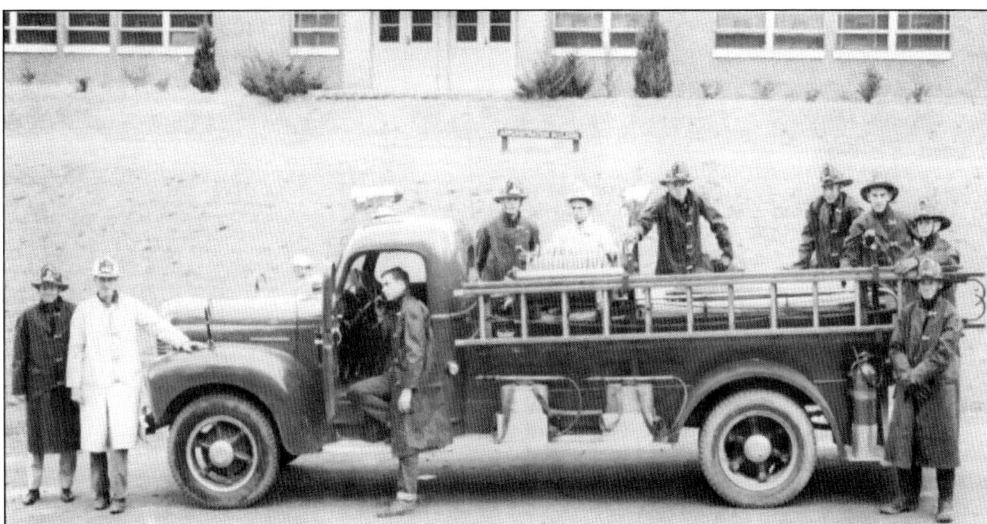

The first truck was wrecked on a call, so ENKA Fire Department built Pisgah this second 250-gallon Army surplus 1941 International truck in 1957. Pisgah responded with ENKA on calls, which would come in to the ENKA plant and then be called into the Pisgah switchboard. The switchboard would activate a siren on the switchboard building and lights and siren in the boys' dorm. The siren would indicate an on- or off-campus call. (Courtesy of Mount Pisgah Academy.)

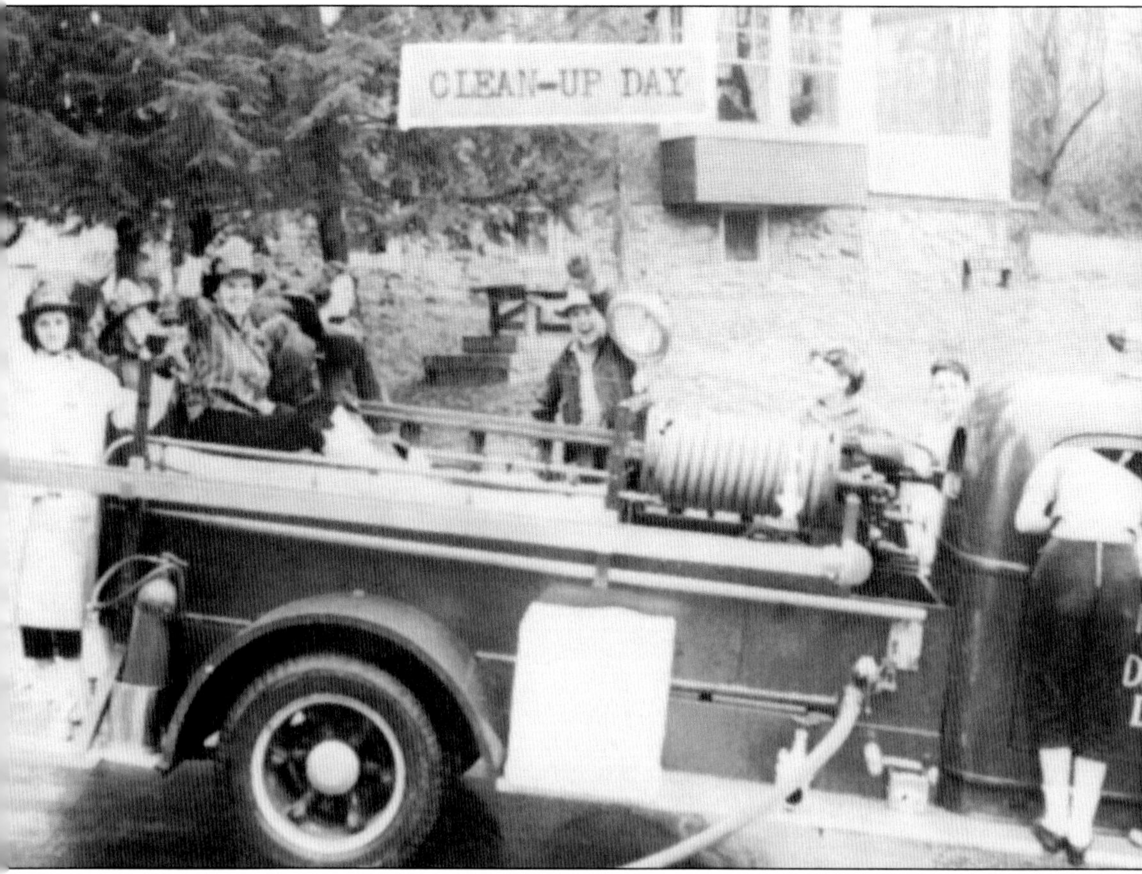

The fire department was very well received in the community. Ranks for the students included captain, first lieutenant, first sergeant, sergeant, mechanic sergeant, and private. The most coveted spots were those of first and second driver. If a student missed class for a fire call, the teachers would average their grade. The fire hall had a single bay and a phone to call the switchboard. (Courtesy of Mount Pisgah Academy.)

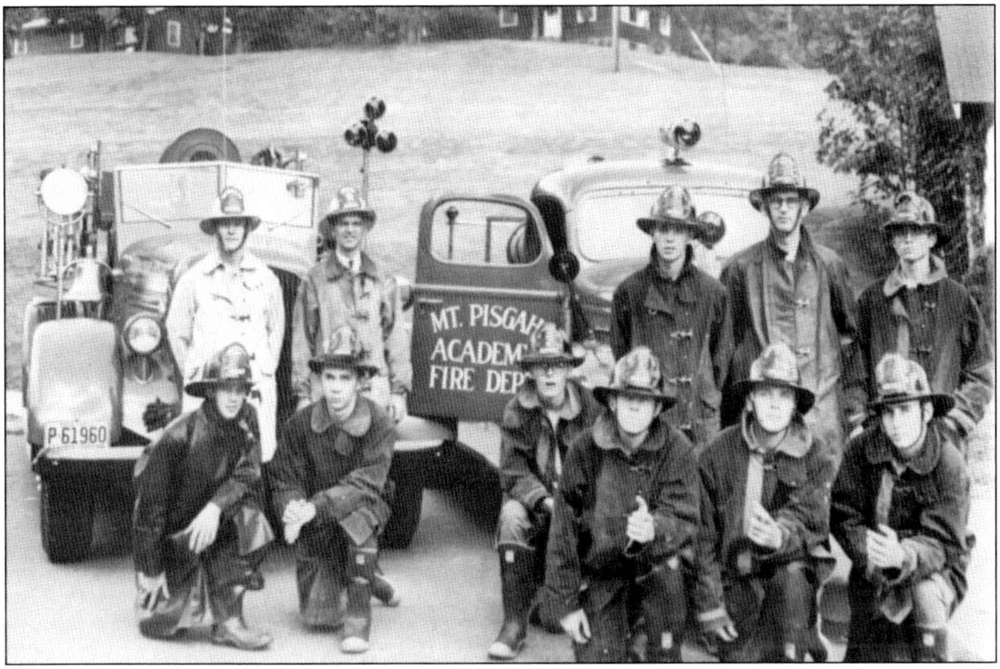

The truck on the left was donated by Tacoma Park, Maryland. The biggest rash of fire calls—14 alarms—came in February 1958. The worst call was a house and woods fire in 1958 during which a man went back inside to get money under his mattress and perished. When insurance costs became too high and Edward Reifsnyder retired, the fire department ended in 1964. (Courtesy of Mount Pisgah Academy.)

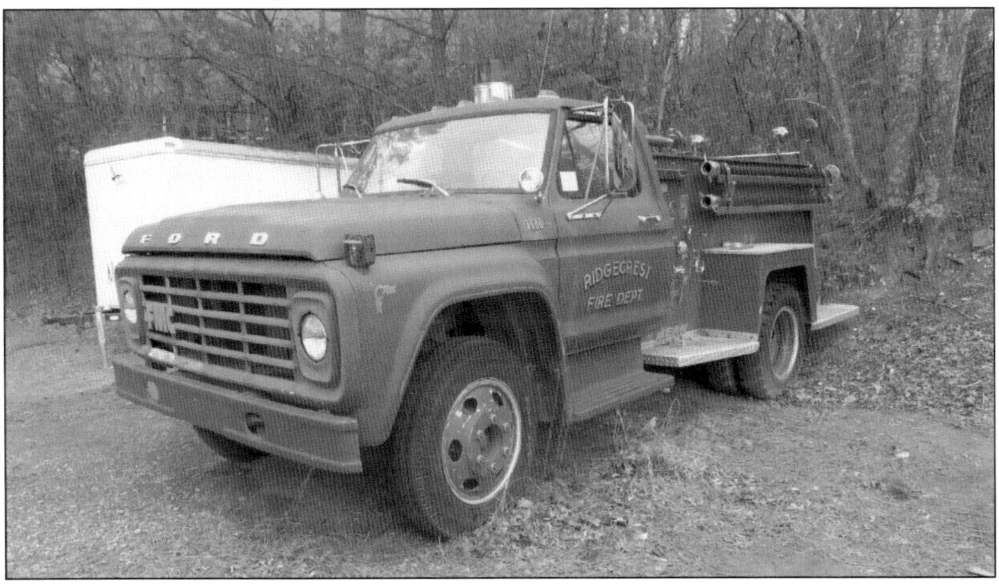

Ridgecrest Conference Center had a fire department staffed by members of the security and maintenance departments and summer staff. A 1957 Chevy pickup with a tank and pump in the truck bed was their first truck. Later, a 1978 Ford FMC with a 300-gallon-per-minute pump and 150-gallon tank was purchased. The fire truck participated in an annual holiday parade held on the grounds. (Author's collection.)

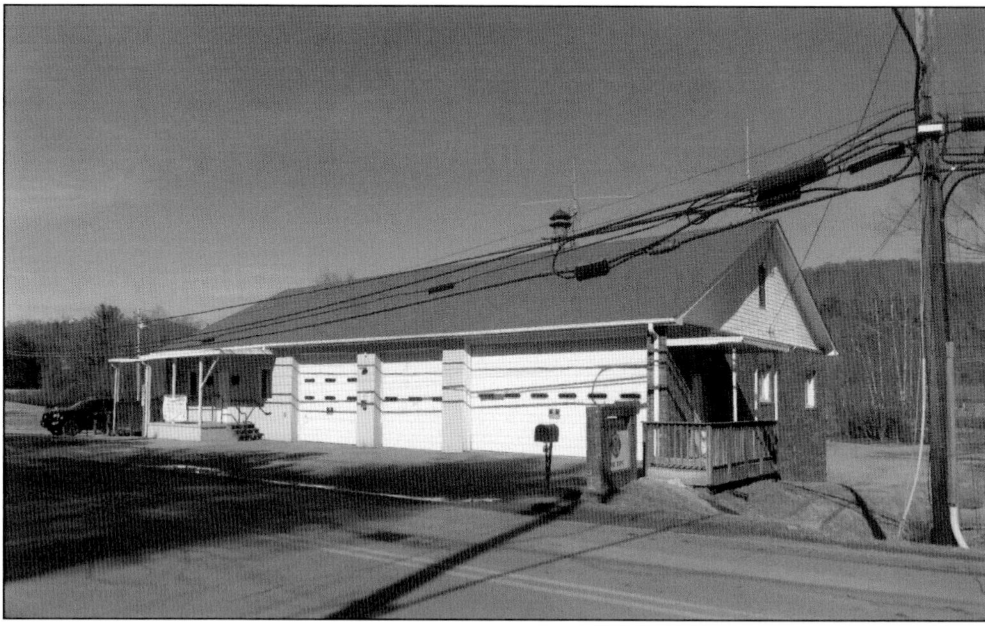

Beaverdam community residents on January 29, 1960, elected a board of directors and agreed to build a combined community center and fire station on land donated by J.R. Stradley for the Beaverdam Volunteer Fire Department. James Carson was elected the first chief. On July 30, 1960, a dedication was held for the new building and new $12,000 American LaFrance fire truck. (Author's collection.)

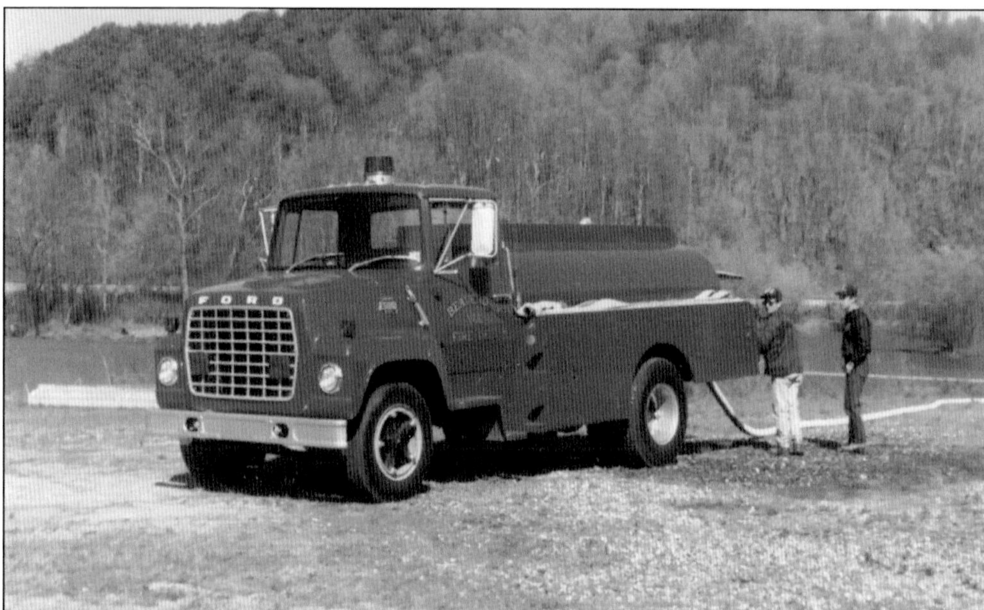

In April 1963, they purchased a Civil Defense (CD) truck and started a rescue squad with Howard Collins as the chief of rescue. On November 13, 1980, a 50-acre wildland fire occurred on the south side of Elk Mountain, threatening 12 homes. Every county fire department responded, and the blaze was contained within a few hours. The department closed and turned operations over to Reems Creek Fire Department on July 1, 1990. (Courtesy of Woodfin Fire Department.)

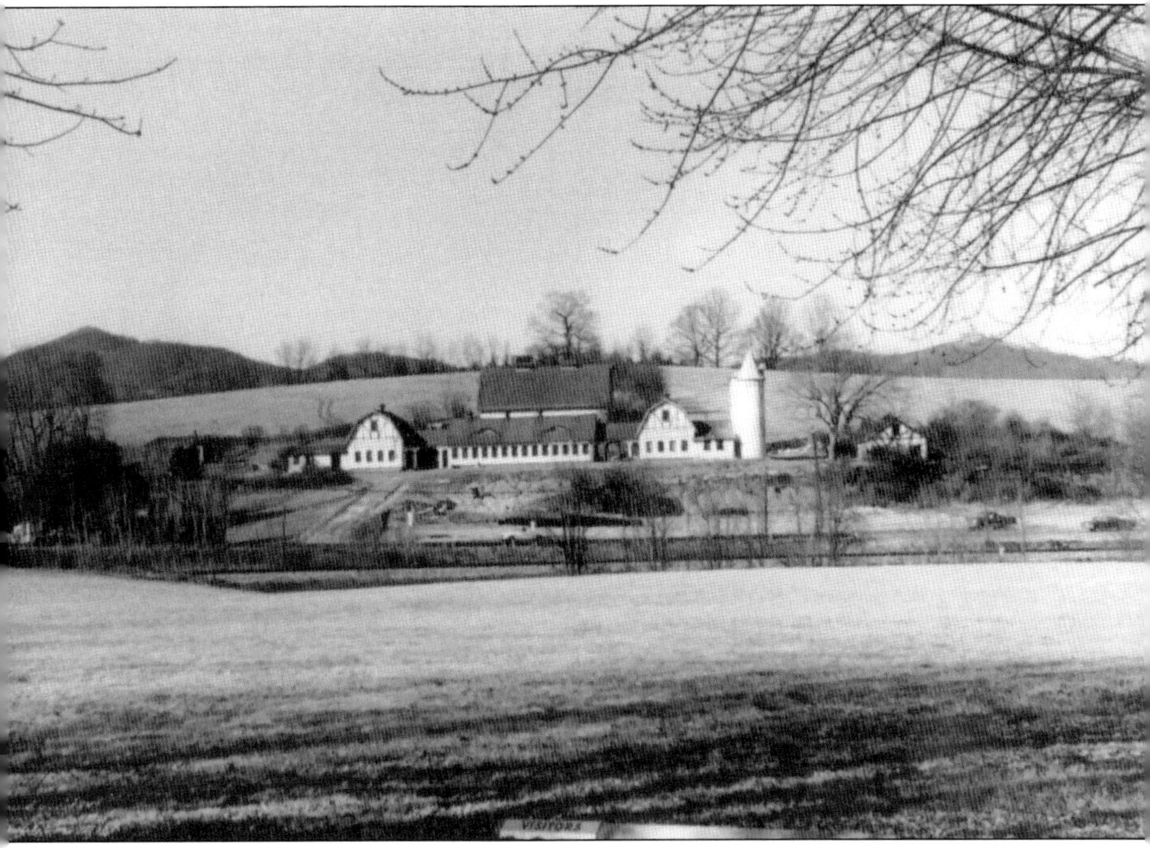

Valkyrie Dairy was located across Smokey Park Highway from the ENKA plant in Candler, about where the Westridge Shopping Center is now. At one time, it had a herd of 40 purebred Guernsey cows. In February 1961, a meeting was held at the Candler School to organize the Candler Volunteer Fire Department, and a board was named. Members of the Pole Creek community were looking to organize a fire department after long travel distances from ENKA Fire Department resulted in a few house fires being too far gone by the time responders arrived. The new department was chartered on March 22, 1961. Another meeting was held in May, and Charles Fletcher was elected chief. A site across from Trantham's grocery store at the end of Dogwood Road was selected for a station, but, lacking enough community support, the fire department never got off the ground. (Courtesy of North Carolina Collection, Pack Memorial Public Library, Asheville.)

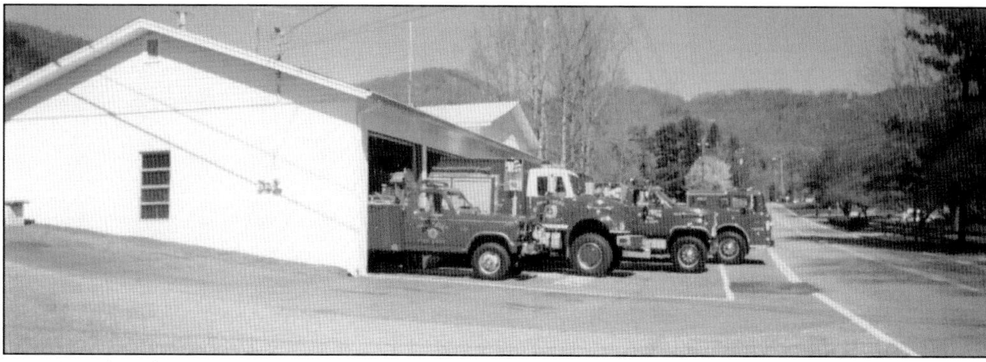

Haw Creek Volunteer Fire Department was founded 1961. The original station was built with two bays and a kitchen; later, a two-story addition was built to the right, two more bays were added to the left, another bay was constructed, and finally, a shop was constructed to the rear. On July 1, 2009, because of a lack of manpower and a shrinking fire district, the Asheville Fire Department took over operations, and the station became AFD Station 12. (Courtesy of John Mulcahey.)

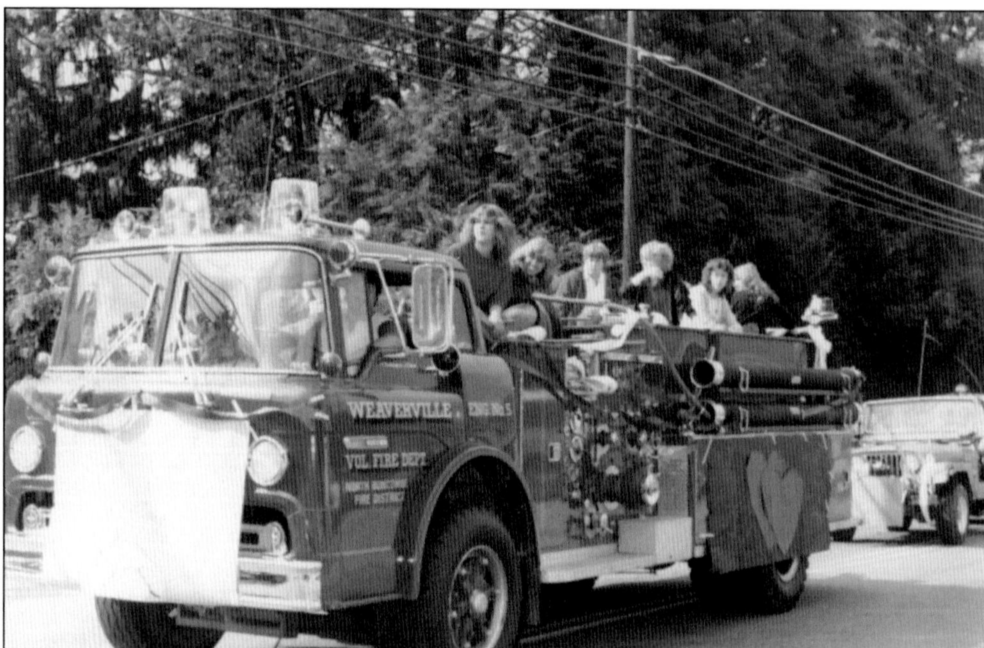

A vote was held on March 23, 1976, for a fire tax in the area known as the North Buncombe Fire District outside the town of Weaverville. A separate board was formed that directed the district's funds until 2000. The North Buncombe Volunteer Fire Department existed within the Weaverville Fire Department. The district bought Weaverville Fire Department the above truck, shown here in the 1982 Fourth of July parade. (Courtesy of Weaverville Fire Department.)

Four

Supporting Agencies

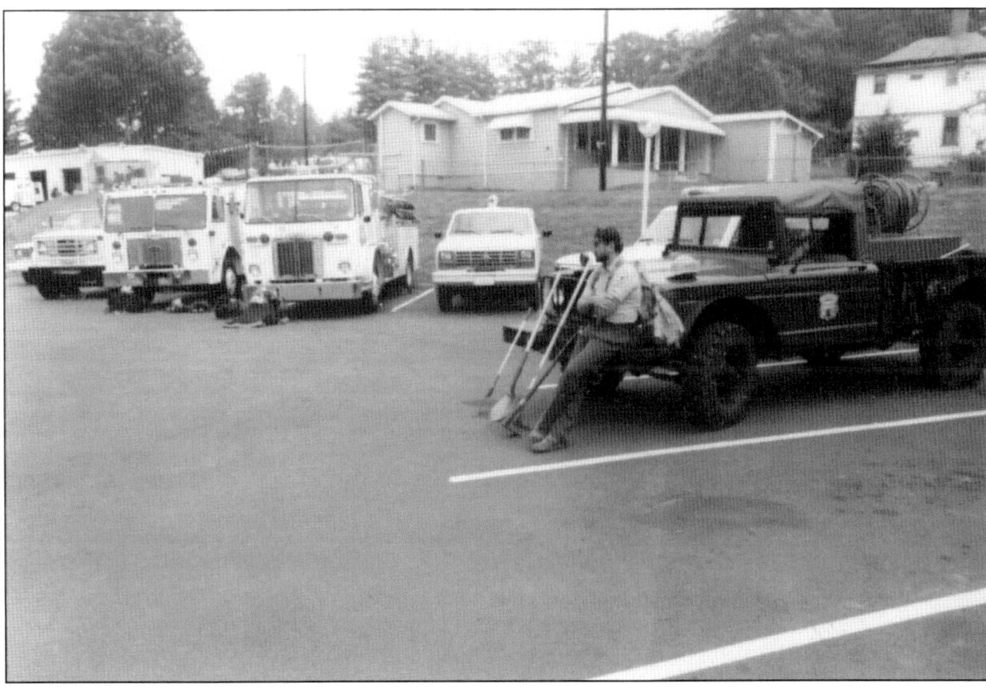

In 1892, Gifford Pinchot launched the nation's first experiment in forestry management on George Vanderbilt's Biltmore Estate. The Biltmore Forest School was started by Dr. Carl Schenck to train the nation's first foresters and included instruction in fire management. In 1921, Buncombe County became one of the first in the state to appropriate money for county wildfire control. The state was divided into districts, each having a district forester. Buncombe County was District 1. (Courtesy of Skyland Fire Department.)

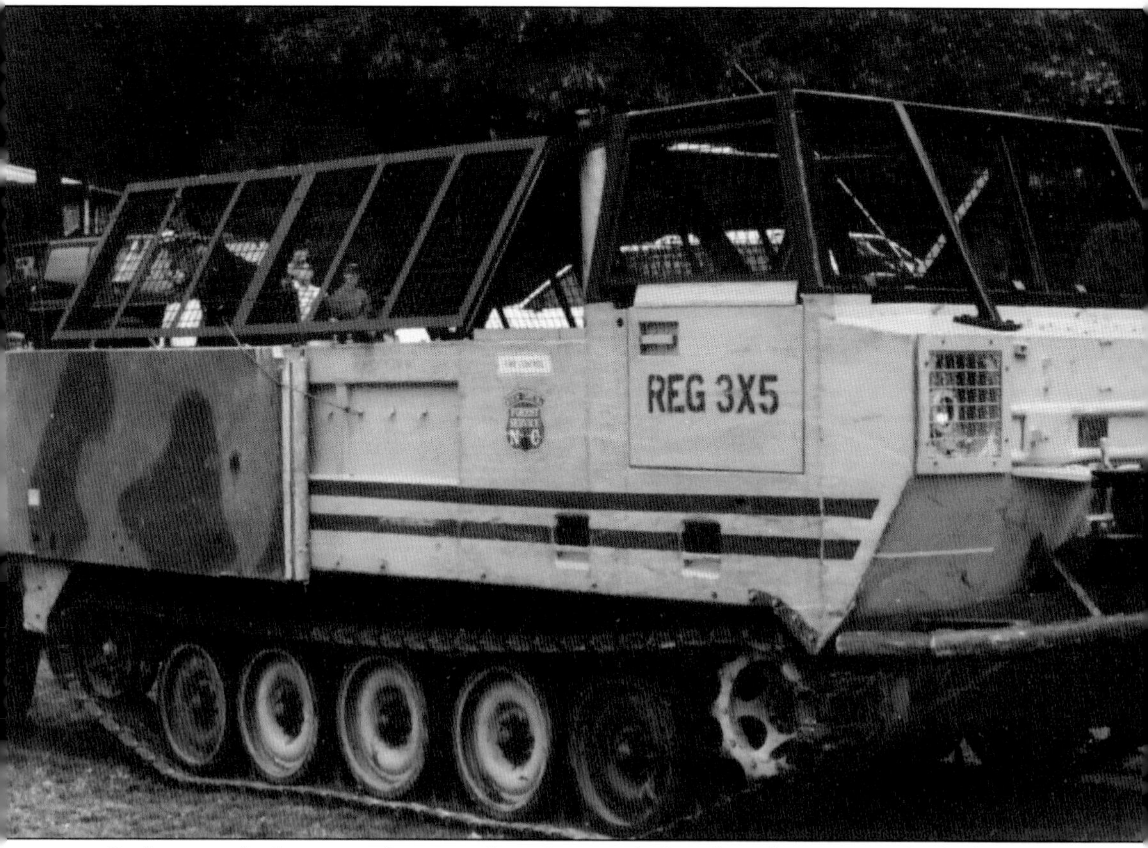

Each county had a county forest warden who issued all tools and equipment for firefighting and kept them in good condition; attended fires both day and night, often going days with no sleep; was responsible for reports on damaged land and timber; and did fire prevention. Equipment included knapsacks, brush hooks, fire flaps, rakes, water buckets, axes, and tractors with plows. Pictured here is some more-modern equipment. (Courtesy of Swannanoa Fire Department.)

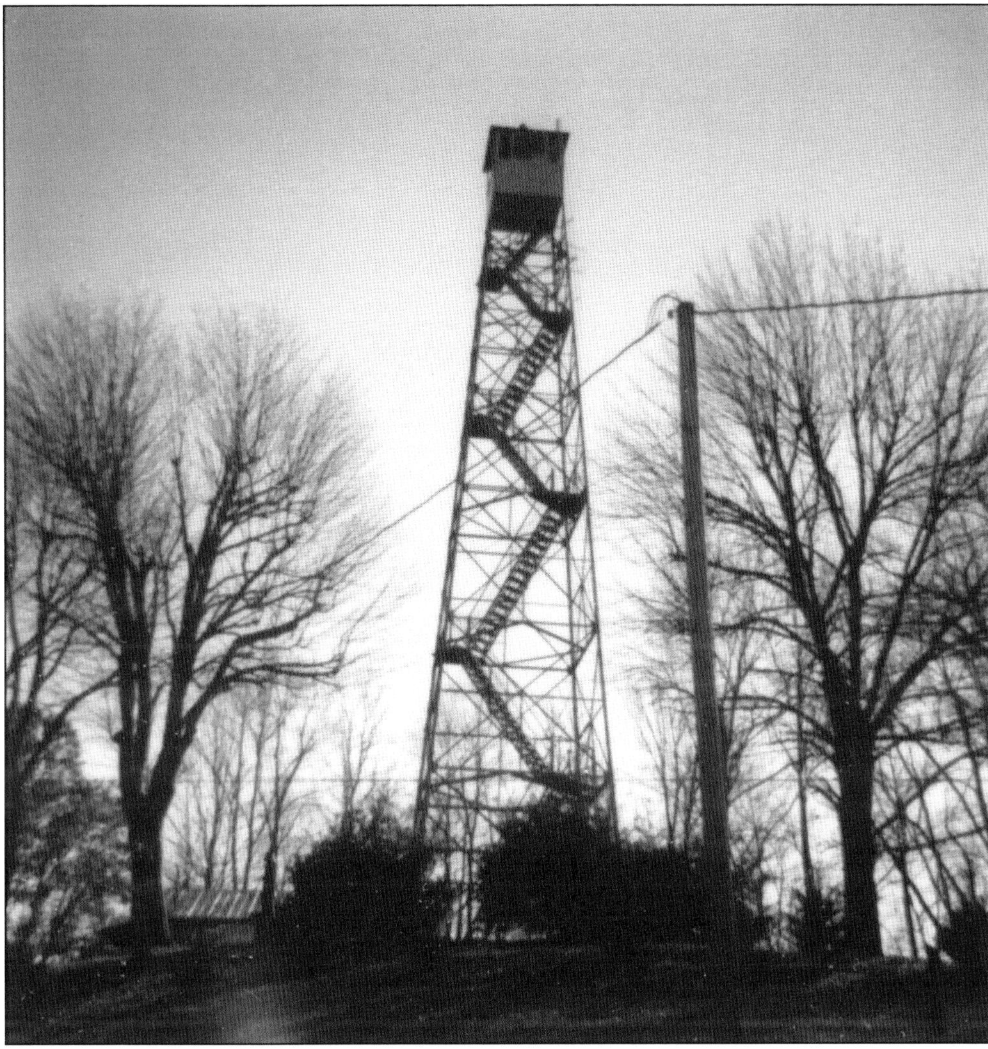

In 1934, the 55-foot-tall Bearwallow fire tower was built on the Buncombe/Henderson County line. It had a two-room cabin beside it and was equipped with a two-way radio instead of the old phone system. If smoke was spotted, other towers were used to triangulate the location with a compass point and map and report it to wardens below. From October to April, during peak wildfire season, towermen and their families would stay on the mountain in the provided cabins. At one time, Bearwallow was the only year-round tower. Attendants at other towers were paid $700 to $800 for a seven-month stay. Several state and federal towers have watched over Buncombe County over the years, including the state-owned Big Nob (shown above, now gone), federal tower Snowball Mountain (now at Big Ivy Community Center), federal tower Duckett Top (still in existence), state-owned Bearwallow (still in existence), federal tower Rich Mountain (still in existence), state-owned Spivey Mountain (now at Holmes Education Forest), federal tower Frying Pan (still in existence), federal tower Green Knob (still in existence), and state-owned High Windy (now gone). Cell phones started the demise of fire towers in the 1990s. (Courtesy of Barnardsville Fire Department.)

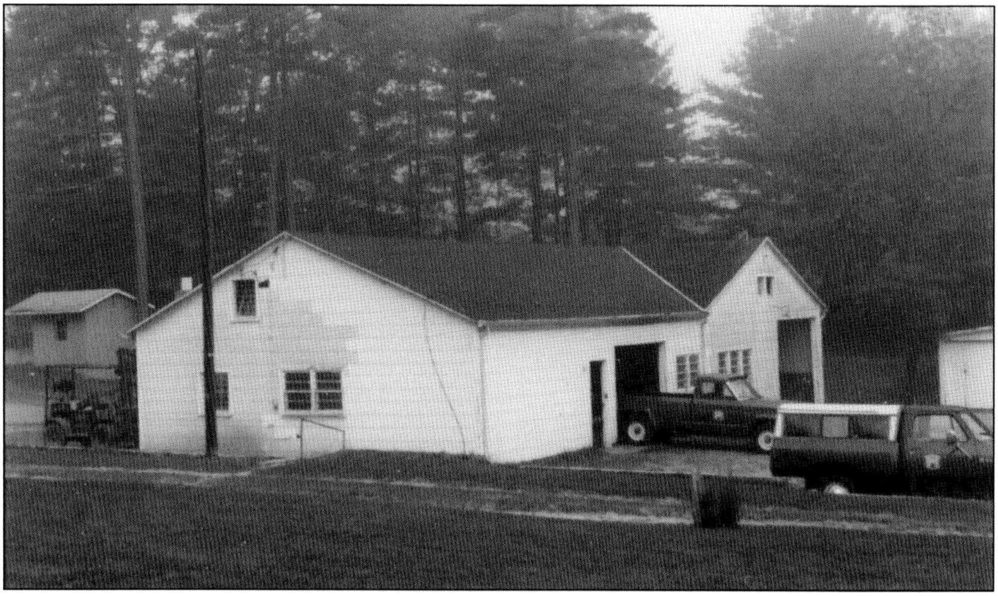

In 1948, district forester B.H. Corpening had his office in the Federal Building in Asheville. The structure above was an old Civilian Conservation Corps building brought down from Mount Mitchell in the 1960s to be used as the District 1 shop. It was torn down in 1987 for the construction of the current building. The Buncombe County Rangers used a small building at the District 1 office before moving into the former construction trailer below, which they occupied 1982 to 2007. It had a wood-burning stove as its heat source, and many a tale was told around it. When the trailer was condemned, they used Skyland Fire Department's Avery Creek station before moving into a new building behind the Juvenile Evaluation Center in 2010. (Above, courtesy of Steve Hensley; below, courtesy of North Carolina Forest Service of Buncombe County.)

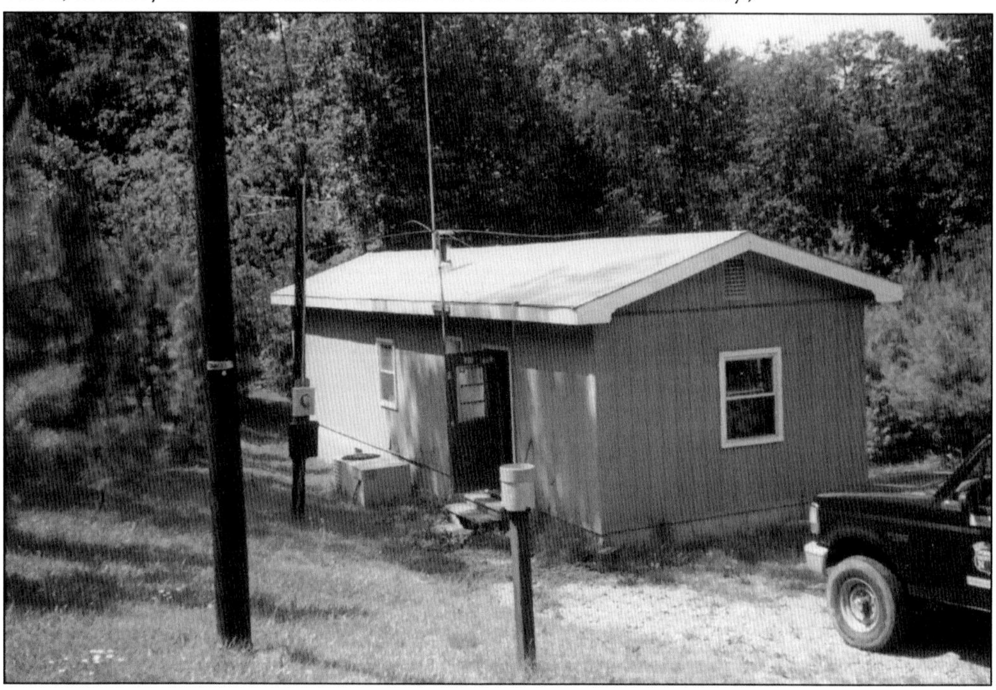

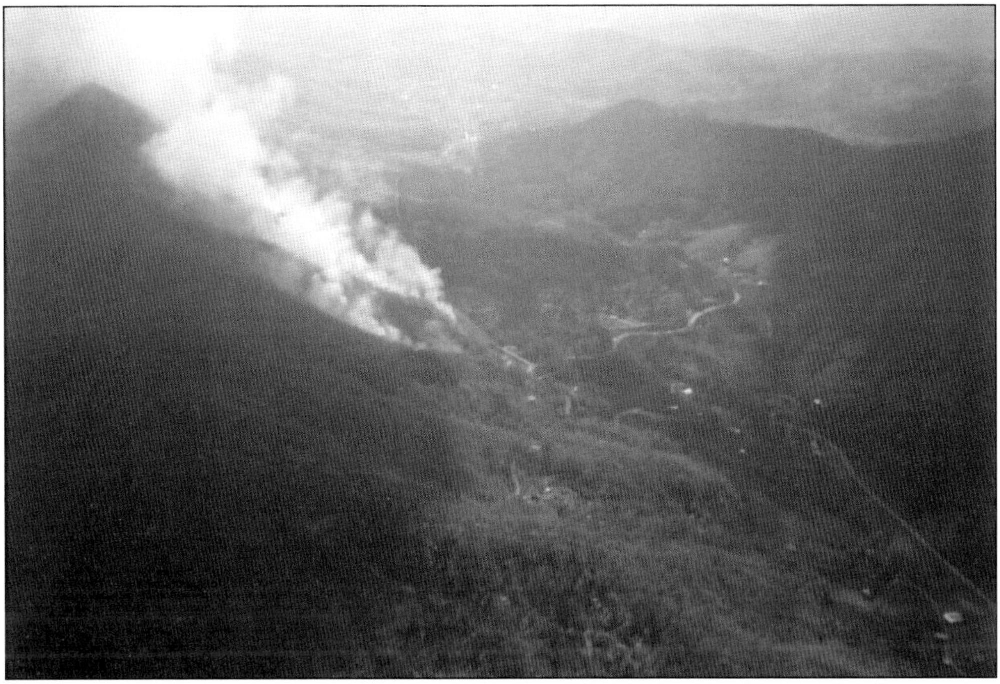

In 1906, National Forest District Rangers were required to be husky; able to ride a horse, rope, shoot, build trails, cook, sleep on the ground, do carpentry work, build a log cabin with an axe, throw the diamond hitch lashing technique, and make topographic maps; possess knowledge of firefighting and entomology; have tact and diplomacy; and be a good businessmen. In the mid-1990s, a fire occurred in Newfound Gap, shown above. The first air tanker, the CL215 shown below, was used. In 1948, the fire-suppression organization included a full-time county forest warden and part-time or seasonal towermen, smokechasers, district wardens, and fire-crew members. (Both, courtesy of Steve Hensley.)

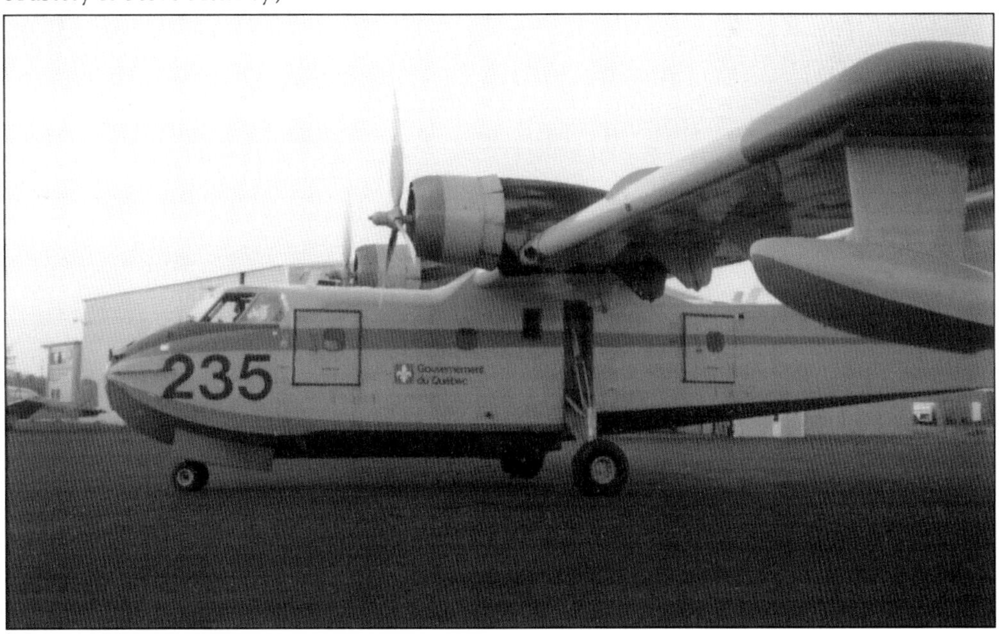

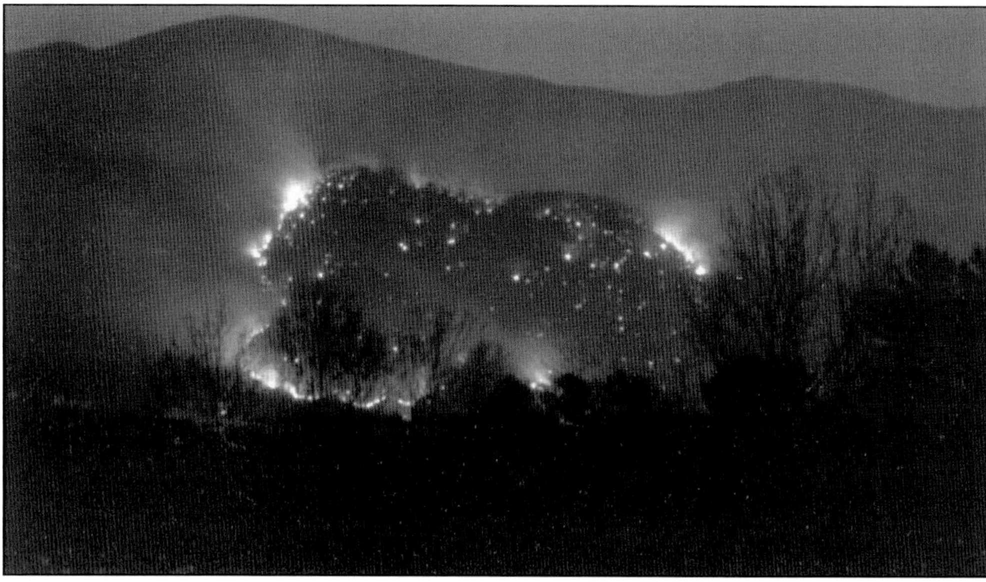

In March 1993, the largest wildland fire in Buncombe County history occurred in Montreat. It covered 700 acres, burned two homes and damaged several others, and threatened 70 homes and a college and conference center. On December 21, 1958, at about 1:15 p.m., 55-year-old Asheville district ranger Thomas Elijah Begley died of a heart attack while fighting a two-acre forest fire started from a trash fire off old Haw Creek Road. He had nearly 20 years on the job and was the assistant in charge of fire control to district forester B.H. Corpening. On August 21, 2008, at about 10:30 a.m., 32-year-old assistant district forester Curtis Jessen fell off the top of Big Bradley Falls while fighting a brush fire and suffered fatal injuries. He started with the forest service in 2002. (Above, courtesy of Steve Hensley; below courtesy of Swannanoa Fire Department.)

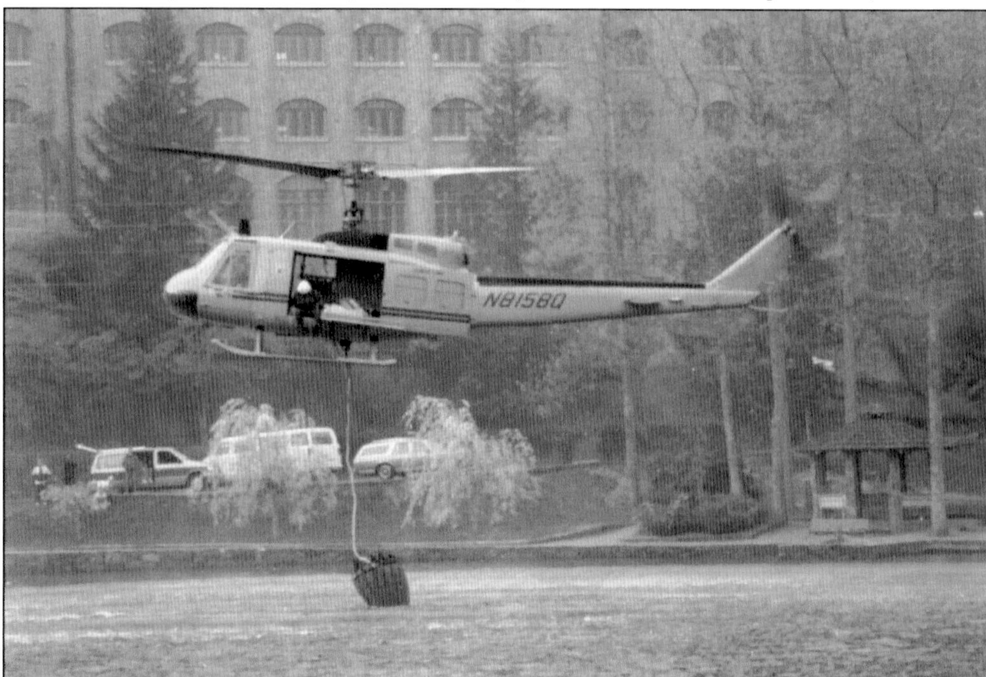

The Buncombe County Rescue Squad was organized on November 2, 1962, as the first countywide squad not attached to a fire department after a number of drowning incidents in the French Broad River and an unsuccessful search for a small boy. Grady Plemmons was the first chief. The squad first met in homes before moving into this old meatpacking house in the spring of 1963. (Courtesy of Buncombe County Rescue Squad.)

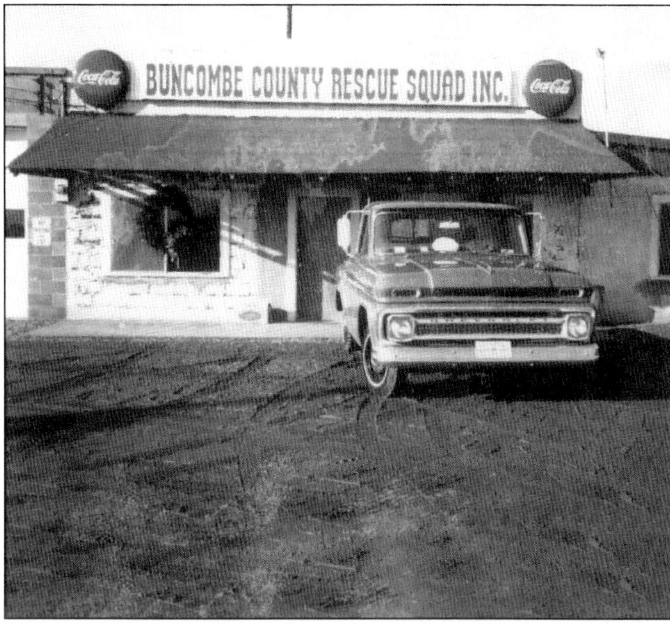

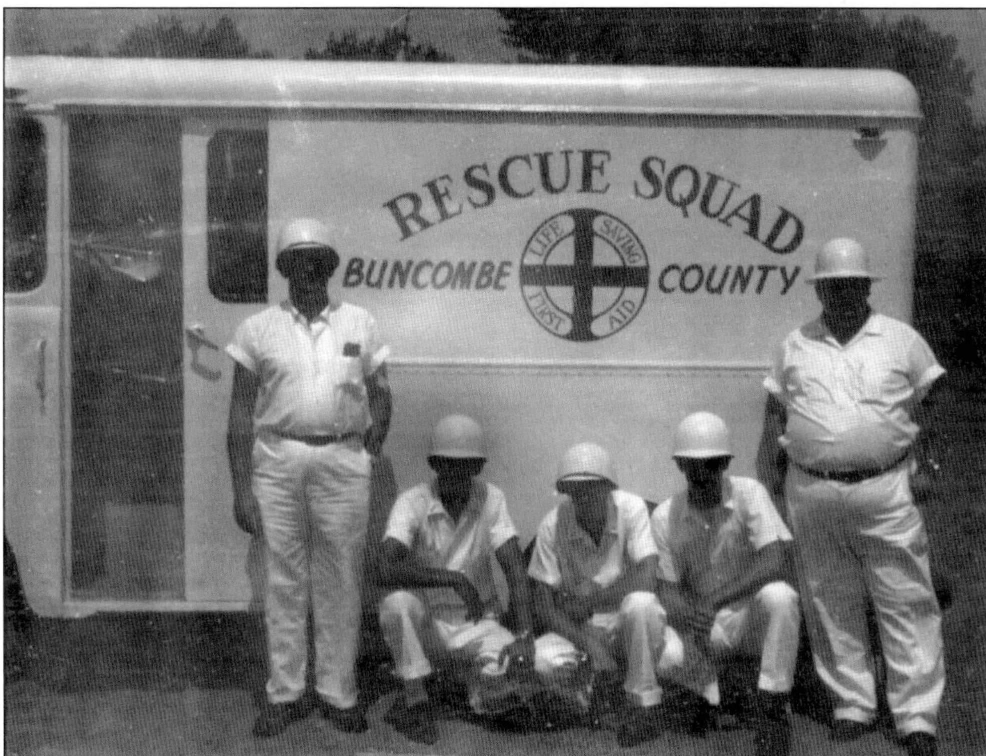

Other rescue squads (with known start dates included) in the area are Beaverdam (April 1963), Black Mountain (January 1962), Fairview, Barnardsville (1964), Weaverville (1961), Skyland (December 1962), West Buncombe, Swannanoa, and Upper Hominy. Each was tied to the civil defense program. Equipment included a set of light-duty rescue tools in addition to drags, ropes, generators, searchlights, and extrication gear. (Courtesy of Buncombe County Rescue Squad.)

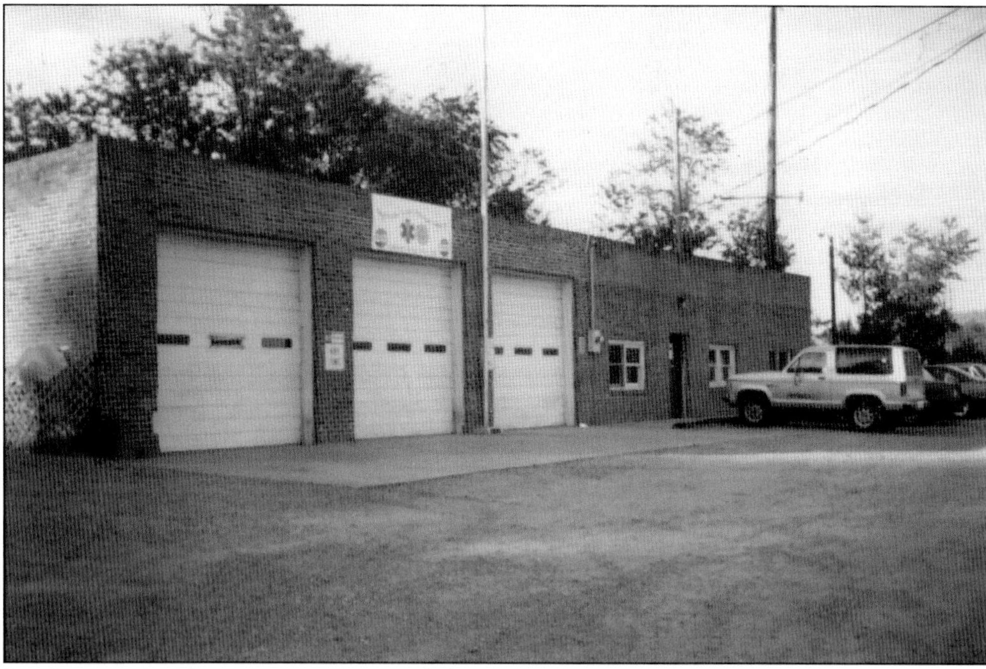

Trucks were first kept in garages around the county before this three-bay addition was made in 1965. At the time, North Carolina was the only state with rescue-service training, having two instructors for 180 squads. Each of four squads had a captain, a lieutenant, and four to six men, and would take turns manning the squad hall on weekends. Equipment included two ambulances, two jeeps, a weapons carrier, a boat, and a pickup truck. (Courtesy of Buncombe County Rescue Squad.)

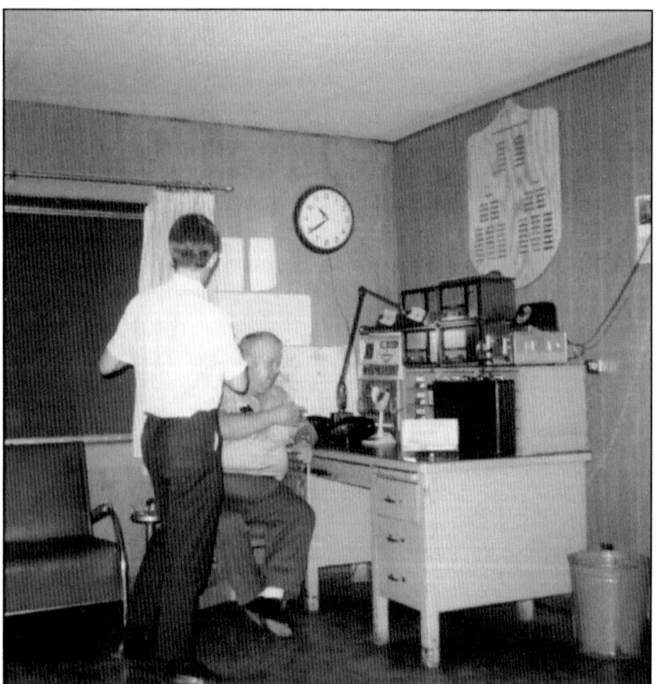

This image shows the squad's radio console. The squad was the first in the state to purchase a Hurst Jaws of Life tool for $3,825. One way they raised money was by using a member's car to block the road and then asking trapped vehicles for donations. The squad house was remodeled and dedicated as the Russell and Thelma Rapp Station on August 28, 2011. (Courtesy of Buncombe County Rescue Squad.)

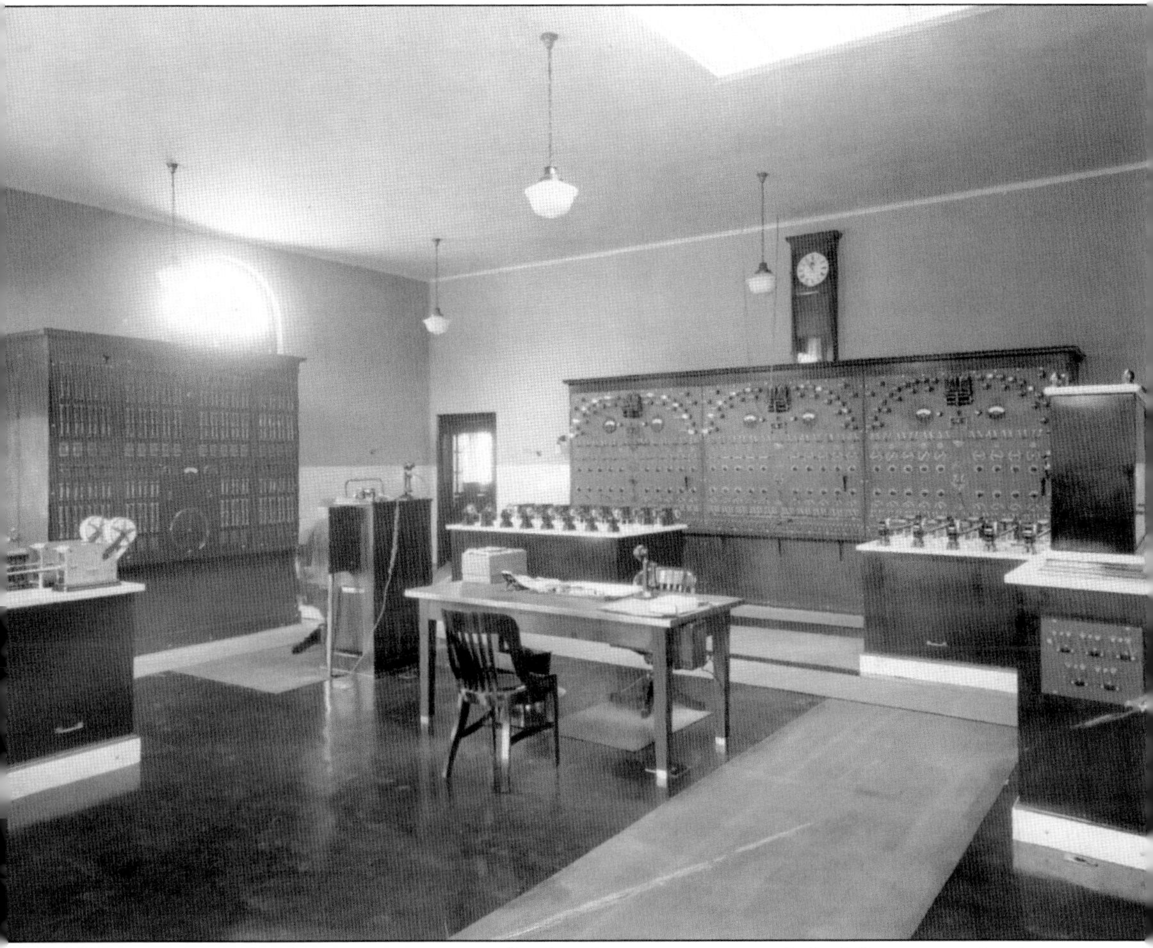

Pull stations were the first way to contact the fire department to get help. The stations had a three-digit number assigned to their location that would ring a town-wide bell to let the firefighters know where to respond. Above is Asheville's alarm receiving station. Volunteer fire departments first operated by telephone. A specified phone number was called by citizens of the fire district, that phone would ring normally or have a continuous ring, the information was gathered by a firefighter or his wife, and then a phone tree would be started calling four other firefighter phone numbers until the entire department was notified. Large sirens on the station were also activated. The address was written on a board at the station for all coming in to see before leaving with the fire truck. Two-way radios were installed in vehicles, and later, walkie-talkies were used to communicate with the fire station and, later, dispatch. A dispatch call center later formed that would activate a speaker called a Plectron, which was kept in members' homes. If no one answered in three minutes, AFD was called to respond. Portable Motorola Minitors came next, allowing a firefighter to be notified everywhere he or she went within the county. (Courtesy of North Carolina Collection, Pack Memorial Public Library, Asheville.)

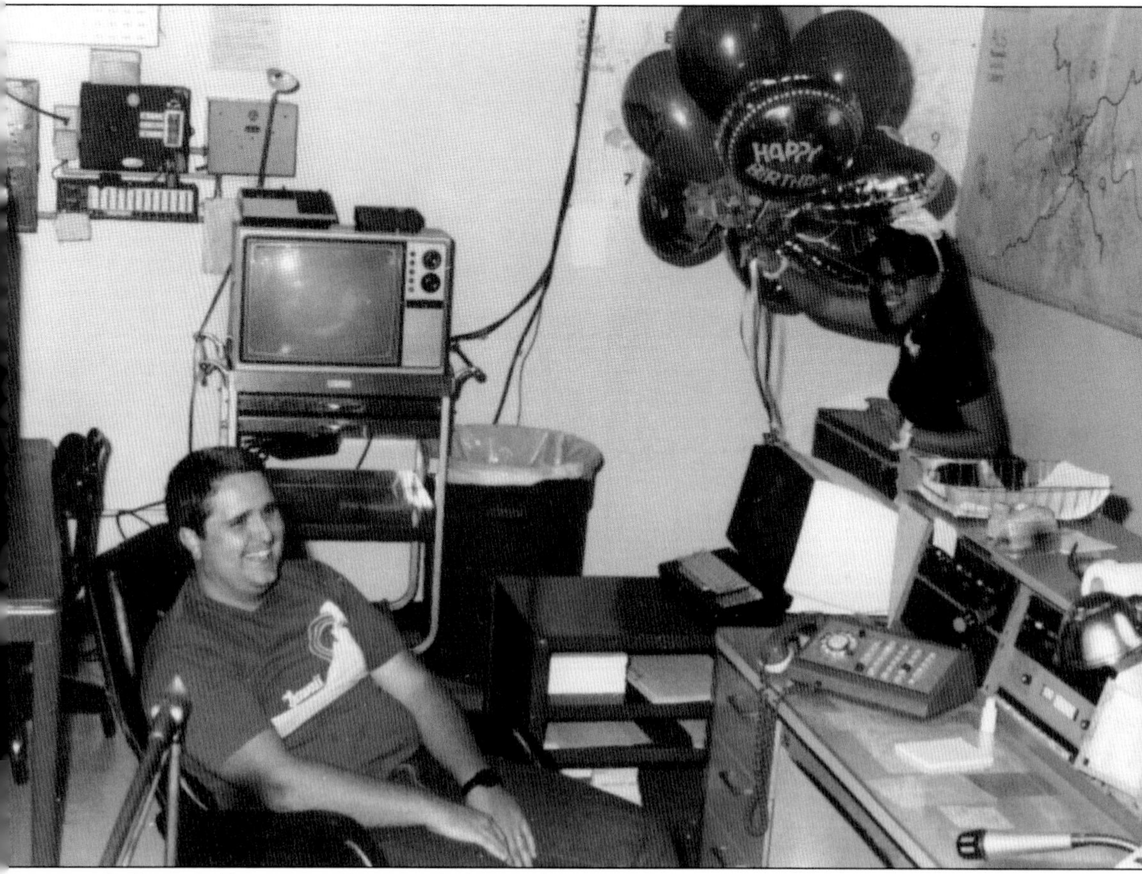

In June 1963, a proposal was made for a radio system for Buncombe County's 18 fire departments with completion by the end of the summer. Dispatch started out in a room on the roof of the Buncombe County Courthouse. The fire departments, sheriff's office, and school buses all shared the same radio frequency. In 1972, dispatch moved to the third floor. It had two EMS and fire dispatchers during the day and one at night that could turn the radio up and sleep on a cot. In September 1975, they moved into the new Emergency Operations Center in the basement of the 1970 Buncombe County Health and Social Services building. They also started dispatching the Buncombe County sheriff's office. Sometime in the 1970s, fire departments were assigned their station numbers because the county ran out of radio numbers. Numbers were picked based on the county ambulance unit number assigned at that station; if no ambulance was stationed there, numbers were drawn out of a hat. (Courtesy of Skyland Fire Department.)

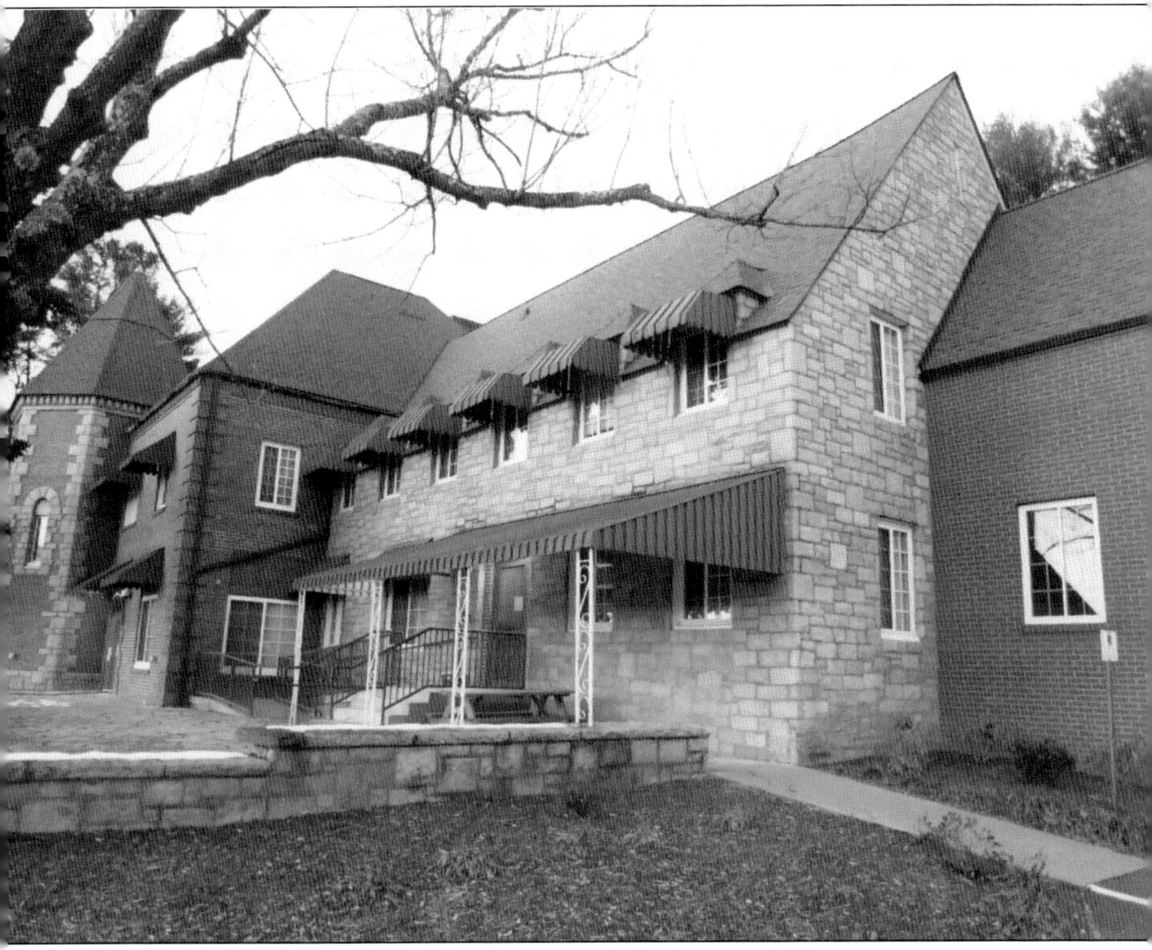

In May 1988, the 911 system went into operation, with the Enhanced 911 system coming a year later. Computer-aided Dispatch (CAD) started in 2003. The city had its own dispatch until the first week of March 2004. In June 2008, the Emergency Operations Center (EOC) moved into a new location, and by December 2009 all county agencies were being dispatched under one roof, including AFD, Buncombe County fire departments (except Black Mountain), Asheville Police Department, Buncombe County Sheriff's Department, and Buncombe County EMS (BCEMS). The new EOC building started out in 1929 as the Buncombe County Training School for Boys. The Buncombe County Home for the Aged moved in 1948 from its nearby 1906 building and remained there until 1975. Then, a private nursing home took over for about five years. It sat empty for a few years before the National Guard put it to use as a training center from April 1983 to 2004. The building went back to the county, and they spent $3.5 million refurbishing it to serve as the new EOC. It is listed in the National Register of Historic Places. (Author's collection.)

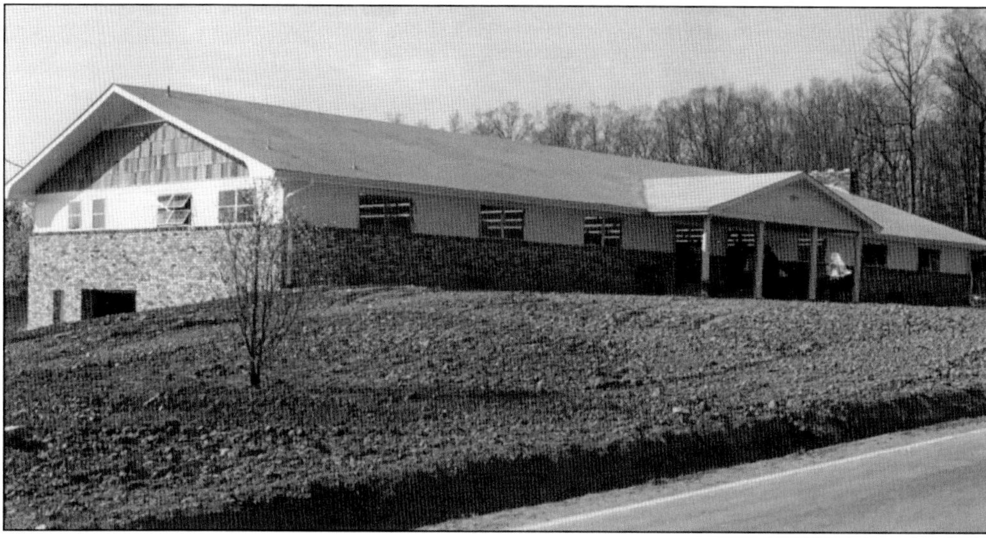

The Buncombe County Firemen's Association formed in January 1962. In October 1965, a decision was made to lease five acres adjoining the Erwin Hills Community Club and the County Home grounds for a training center. The Buncombe County Volunteer Firemen's Association, Incorporated, was chartered in November 1965 with Dick Dillingham from Barnardsville and Ernest Ellis from Woodfin as incorporators and directors. (Courtesy of Barnardsville Fire Department.)

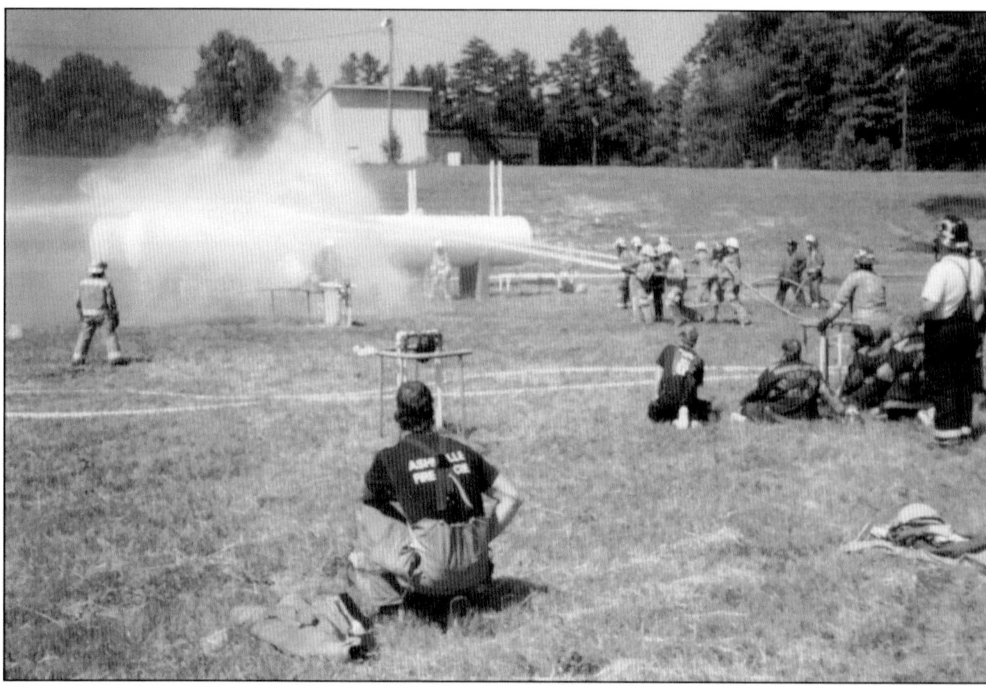

The 12,000-square-foot Buncombe County Volunteer Firemen's Training Center was dedicated on April 7, 1967, complete with a large meeting room, kitchen, offices, and equipment rooms. Other structures followed, including the flammable-liquids and gas pit shown above and the smokehouse seen at the top of the hill. Buncombe County had 17 volunteer fire departments at the time. (Courtesy of West Buncombe Fire Department.)

The training tower was used for exercises such as high-angle rescue (shown here), hose advancement, bail out, Davis-Dodson (firefighter agility) tests, ground ladder deployment, and even a required jump into a safety net. Annual Buncombe County Fire Training Schools, junior firefighter competitions, and fire academies used to be held. A new state-of-the-art training center was opened on old landfill property in October 2012, and all structures except the meeting hall were removed from the old center. (Courtesy of Buncombe County Rescue Squad.)

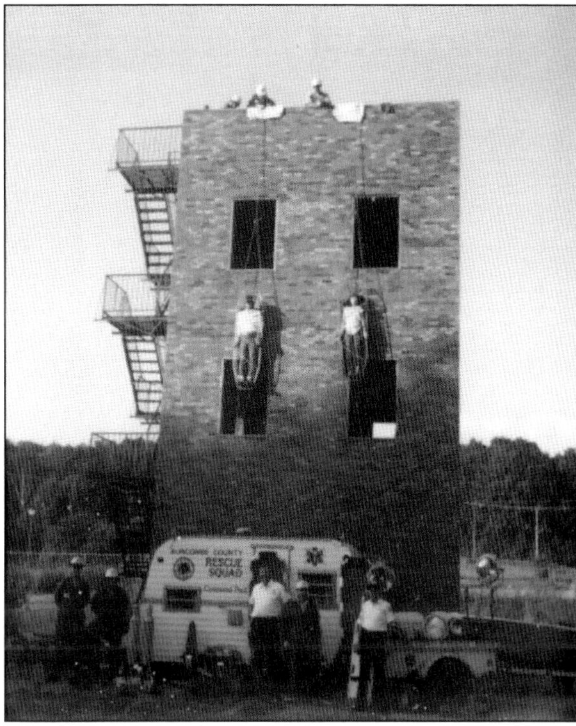

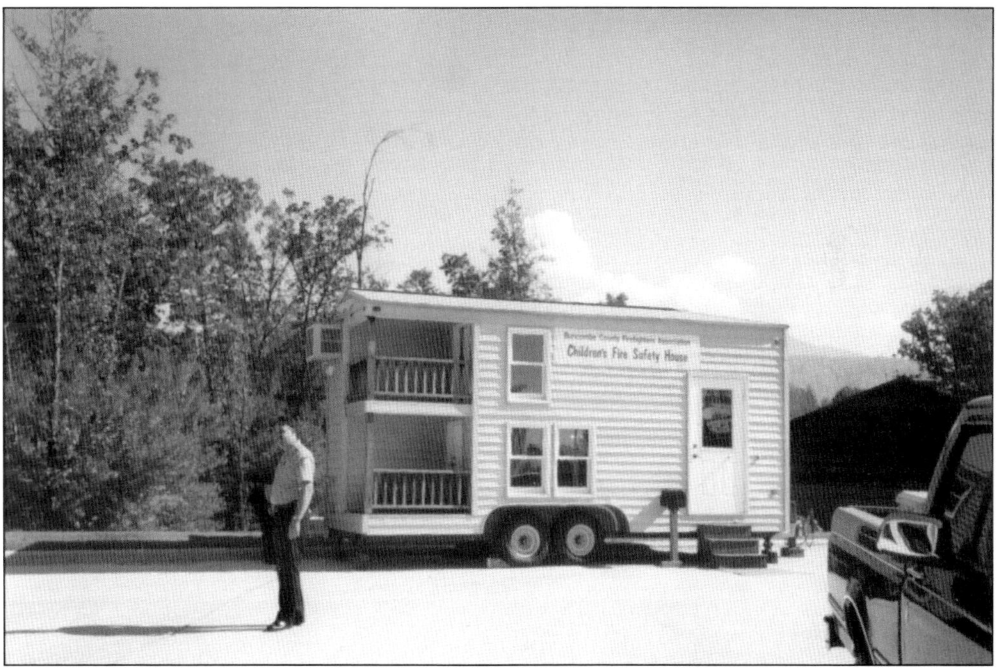

The Buncombe County Firefighters Association built this fire-safety house, complete with living room, kitchen, and upstairs bedroom, to be used by fire departments around the county. It was an invaluable tool to teach children about fire safety. Fire-prevention education also included the life-sized Sparky costume worn by firefighters, puppet shows, skits, truck and station tours, fire-gear donning, and a lot of handouts. (Courtesy of Jupiter Fire Department.)

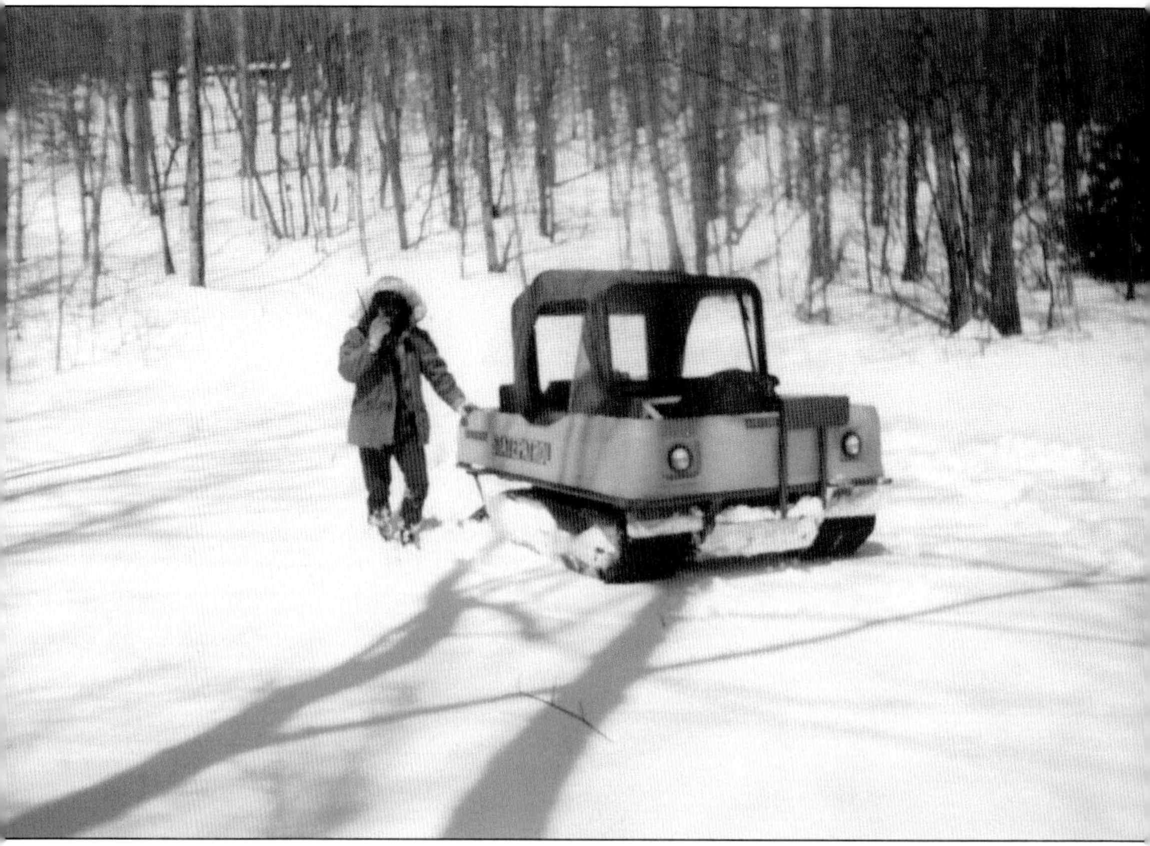

On March 13, 1993, the strongest blizzard of the century—with drifts of up to five feet and winds of over 64 miles per hour—hit Buncombe County, dumping over 18 inches of snow. Flat Top Mountain recorded 101-mile-per-hour winds. The storm knocked out power, water, and telephone service to thousands. The interstate was closed with stranded motorists and minor accidents. The National Guard assisted in reaching those motorists. First responders got stranded responding to calls in their cars and had to abandon them to go in four-wheel-drive vehicles. Area agencies rescued people and took them to safety and also treated the injured or ill, transporting them to medical facilities, but they delayed in doing so by the deep snow. One sad example of the delay was a heart attack patient who expired after it took one hour to reach them in hip-deep snow. Four Asheville and five Buncombe County deaths were caused by the storm or delays in response. Seen above is a member of the Swannanoa Fire Department using a State Highway Patrol vehicle to navigate the snow. (Courtesy of Swannanoa Fire Department.)

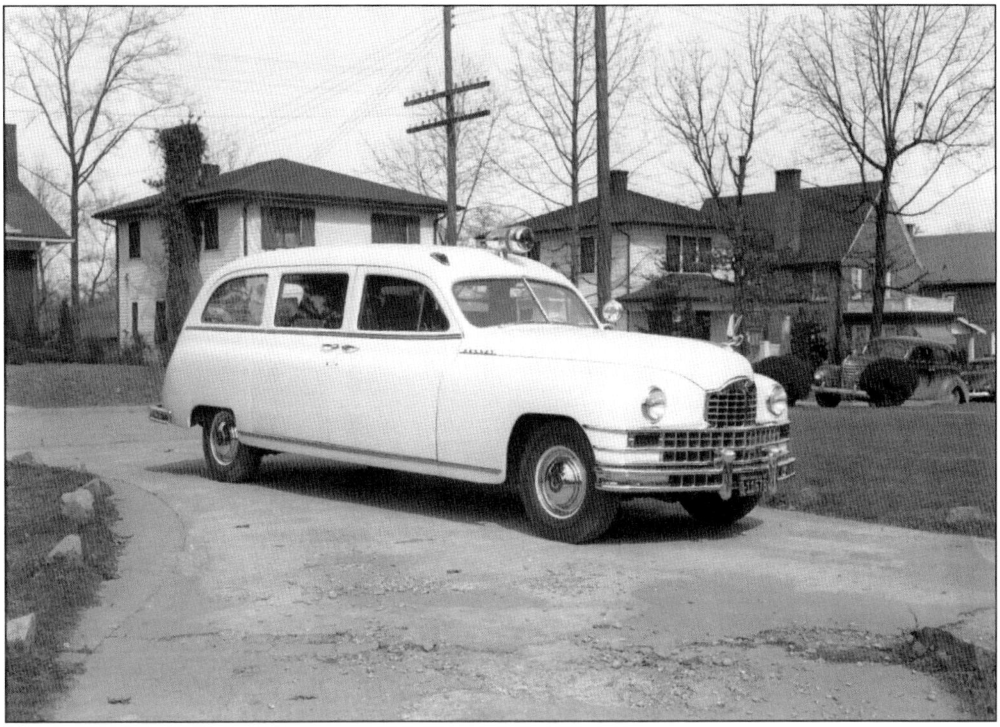

Because funeral homes had vehicles in which someone could lie flat comfortably, they served as ambulances to transport the sick to hospitals. It was truly a load-and-go process, with patients just being put on stretchers and driven off. Groce, Morris, Anders-Rice, Jesse Ray, Allen, West, Penland, and Dunn all provided ambulance services. Harrison in Black Mountain provided an ambulance to the fire department. The ambulances were Cadillacs and Pontiacs, with two personnel. (Courtesy of E.M. Ball Collection, Special Collections, Ramsey Library, UNCA.)

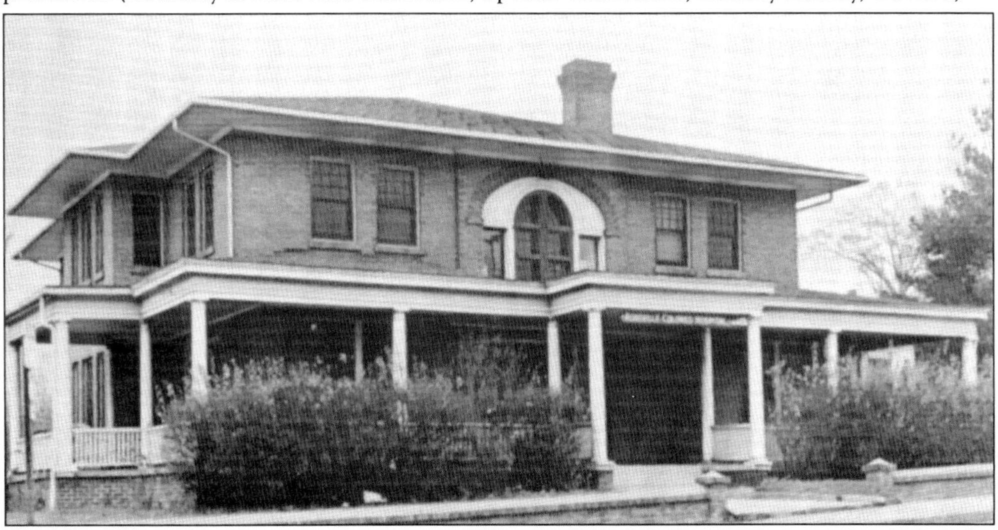

Dr. Mary Frances Shuford opened this building in 1944 for the Asheville Colored Hospital on the corner of Biltmore and Southside Avenues. It later became the Jesse Ray Funeral Home, which had an ambulance service. The building, which still stands today, would go on to become a law office. (Courtesy of North Carolina Collection, Pack Memorial Public Library, Asheville.)

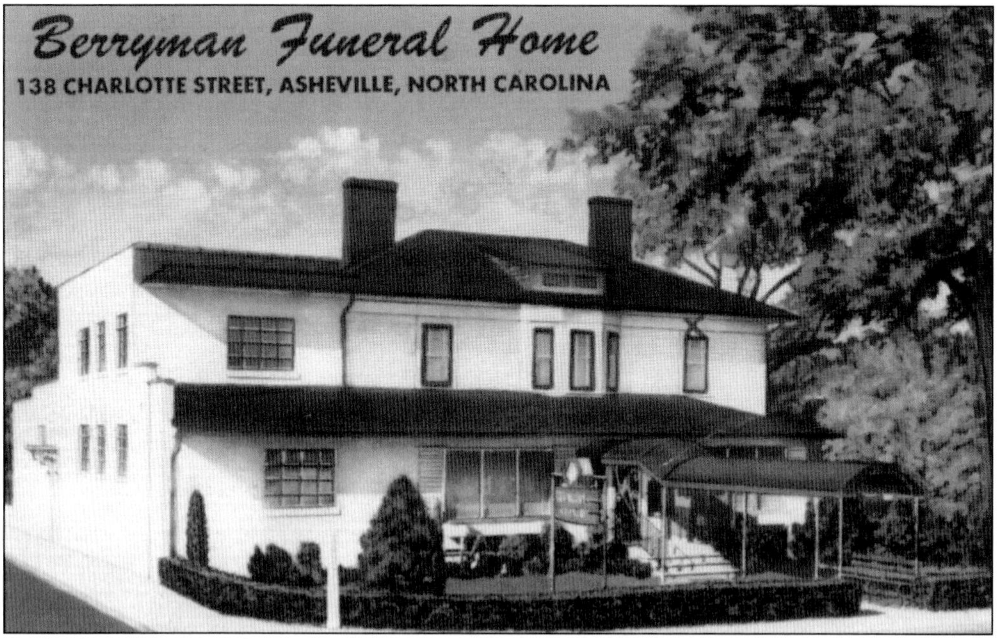

Penland & Sons Funeral Home bought this business in 1968. It already had an ambulance service, which Penland continued to operate (also adding a taxi service). They had up to 10 ambulances with two attendants each. One Cadillac was known as "the Banana." They also had a location in Swannanoa. The building burned down on January 6, 1986. Penland was the last funeral home to operate an ambulance service. (Courtesy of North Carolina Collection, Pack Memorial Public Library, Asheville.)

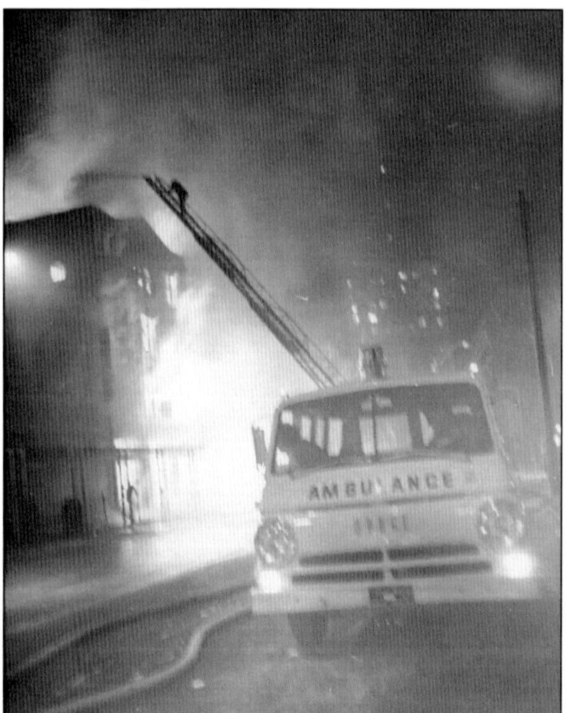

In July 1966, Medicare went into effect, requiring ambulance services to pay a minimum wage of $1.25 an hour, have 24-hour availability, have new training standards, and limit employee hours. Because of the cost, most funeral homes informed the Buncombe County Commissioners that they would no longer provide service after September 10, 1965. Asheville and Buncombe County contracted with West Funeral Home and the new Ambulance Service of Asheville (ASA). (Courtesy of Asheville Fire Department.)

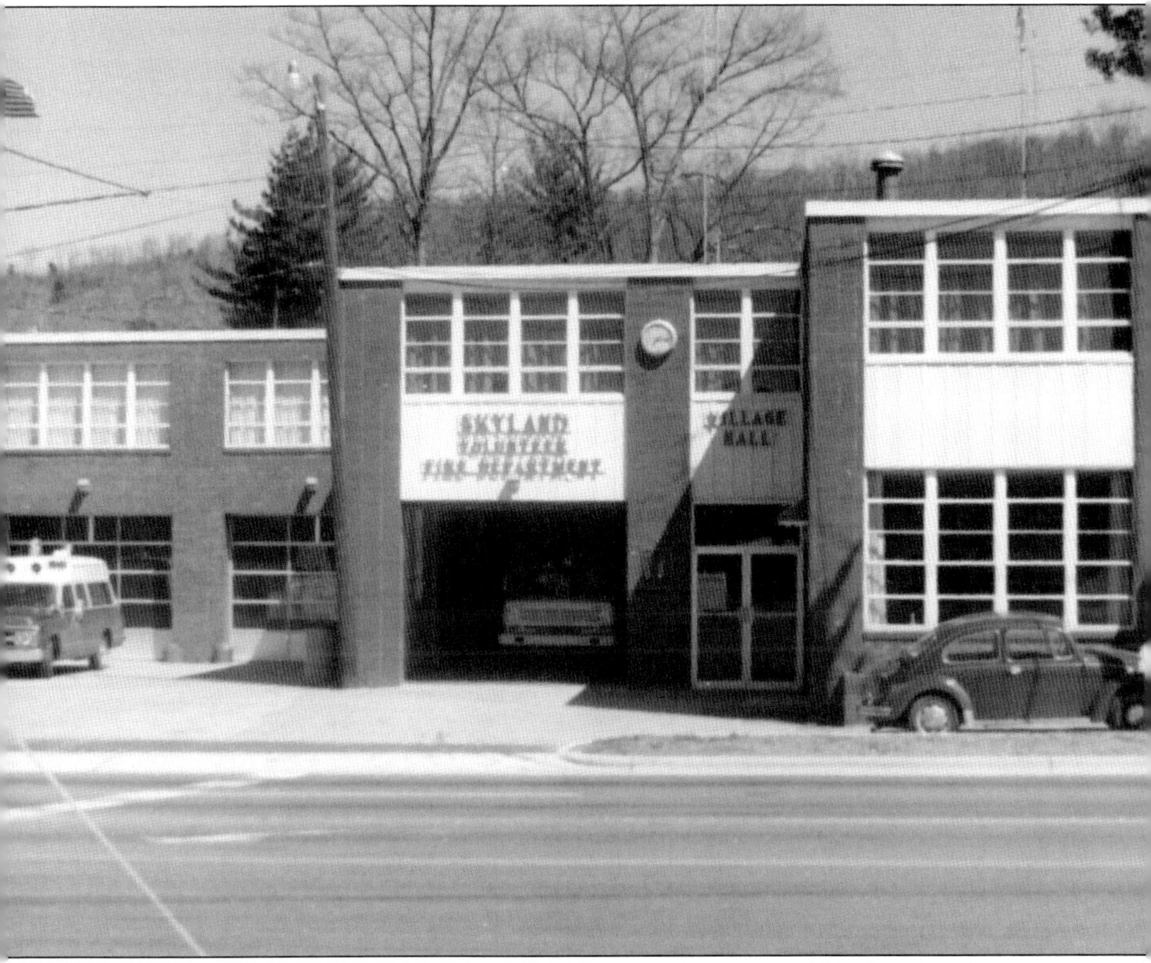

ASA, with firms in Charlotte and Columbia, had their headquarters at 384 Merrimon Avenue and ran a fleet of four new Dodge Sportsman vehicles complete with the newly required two-way radios, proper first-aid equipment, insurance, and personnel trained in first aid. Each ambulance was staffed with a driver and attendant, and all personnel lived at headquarters. Funeral homes in Weaverville, Black Mountain, and two in Asheville continued to operate ambulance services. After nine months, ASA announced they were losing money and could no longer operate. The City of Asheville and Buncombe County joined forces to publicly operate the Asheville-Buncombe County Ambulance Department, headquartered in the former AFD temporary Station 8 on Tunnel Road, starting on September 1, 1966. Jack Leatherwood was the first director. Five new ambulances were bought and stationed, two at headquarters, one at AFD headquarters on Pack Square, one at Skyland Fire Department, and one at AFD Station 6 in West Asheville. A $20 flat rate was charged initially for services, regardless of time spent or miles traveled. (Courtesy of Skyland Fire Department.)

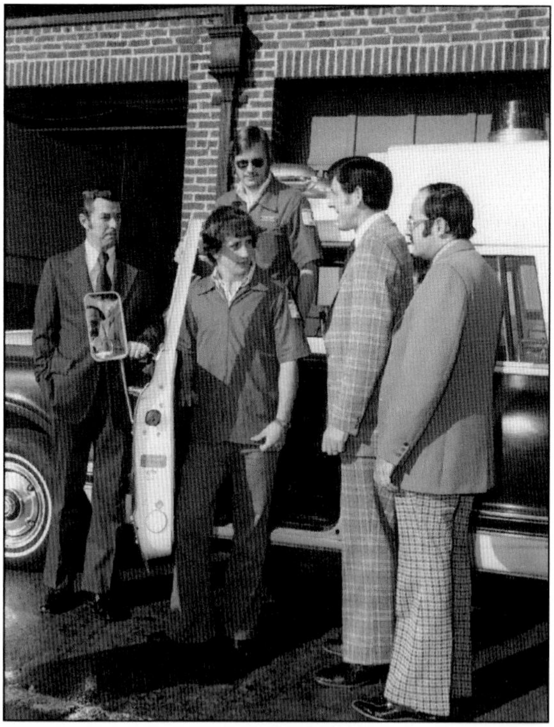

On June 1, 1968, Buncombe County took over sole operation of the Buncombe County Ambulance Service. Emergency Medical Technician training began in January 1974. In February 1975, the ambulance service converted from mostly part time to full-time technicians. EMS 7 (left) at the ENKA plant began operating in March 1975. Darrell Maewether, hired in April 1975, became the first female EMT. (Courtesy of Jerry Vehaun.)

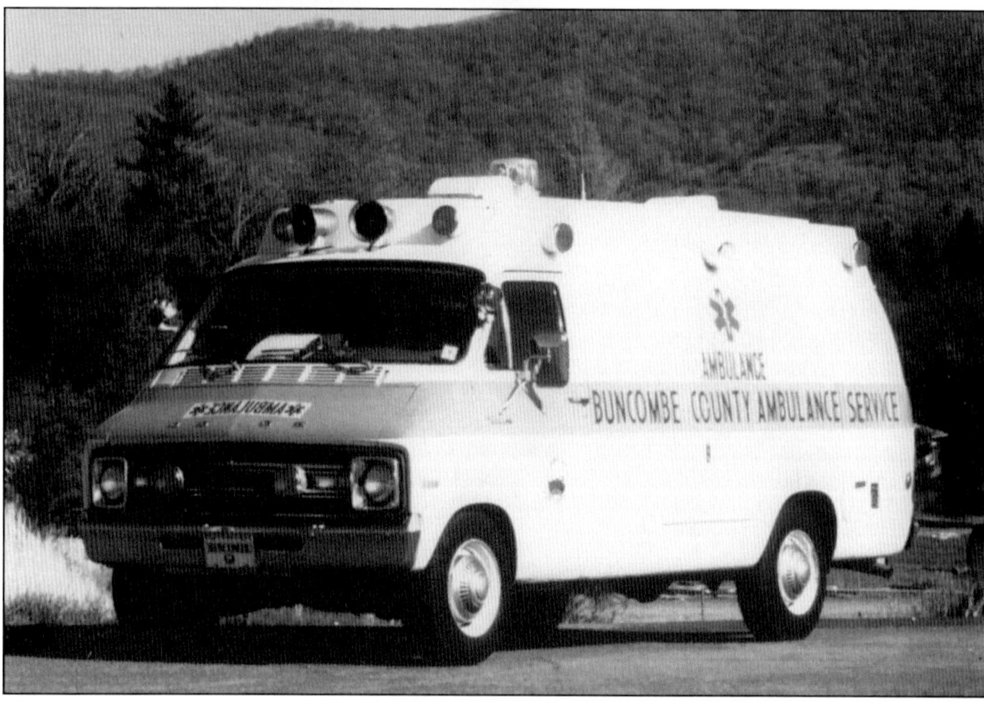

The first ambulances used were the short-van type, followed by wagons like the new EMS 7. Vans started around February 1976, then box trucks in 1977. EMT intermediate classes started in 1977, and the first IV started in the field was in May 1978. The services's name was changed in 1980 to Buncombe County Emergency Medical Service. (Courtesy of Jerry Vehaun.)

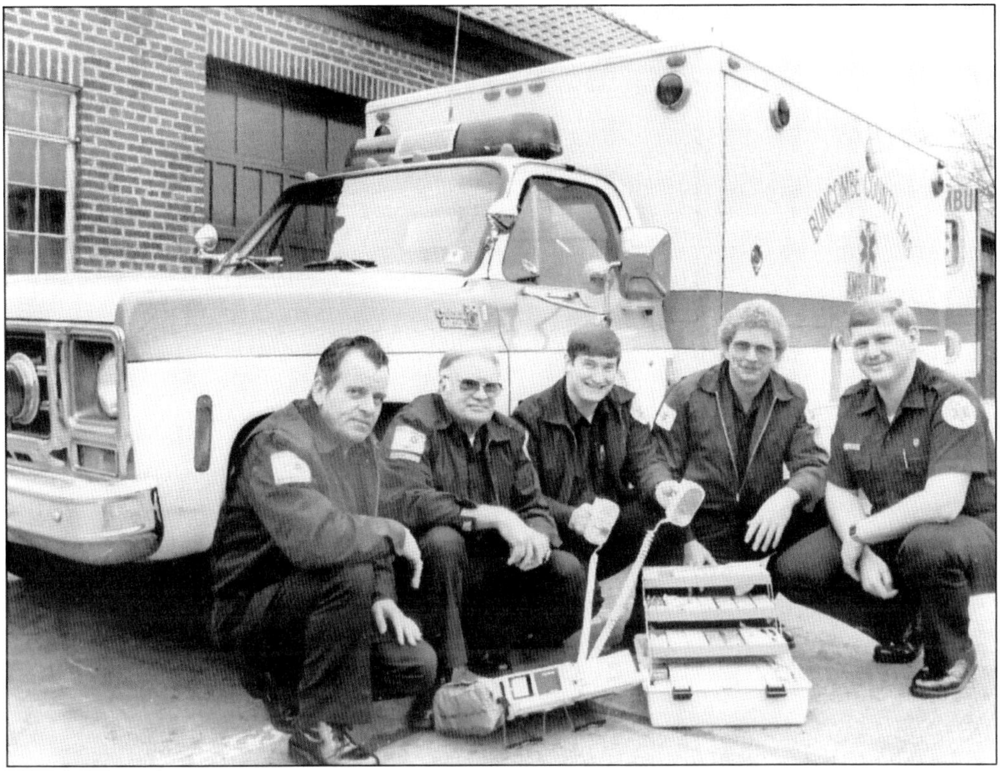

On April 5, 1982, Reynolds Fire Department, Skyland Fire Department, Weaverville Fire Department, and Asheville Fire Department Station 3 ambulances became paramedic units, Black Mountain Fire Department followed suit in 1983, as did the final ENKA Fire Department in 1984. Right after announcing the advancement of 15 new paramedics at a ceremony, two units responded to a gunshot victim—a Wells Fargo armored-car guard—at 11:02 a.m. at K-Mart on Patton Avenue, making this the first paramedic call. (Courtesy of Jerry Vehaun.)

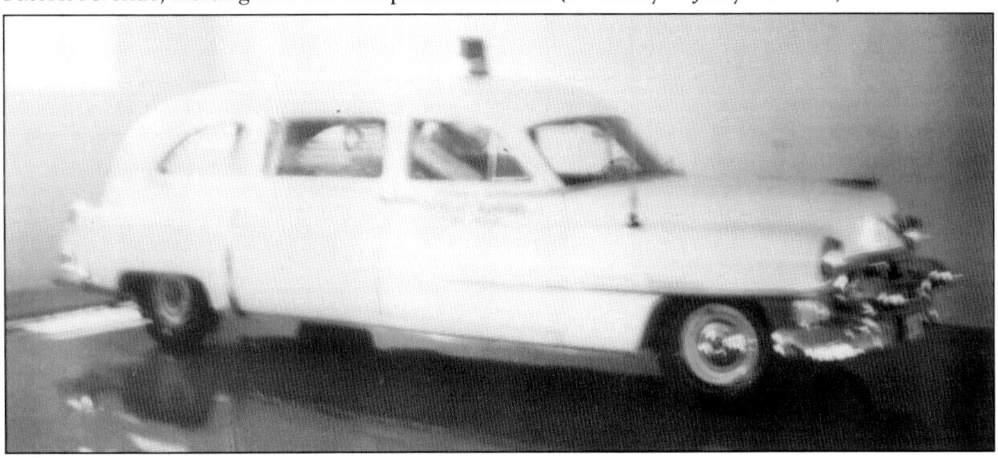

Shown above in 1965 is Riceville Fire Department's first ambulance, a 1953 Cadillac purchased for $609. Ambulance service in the county included funeral homes, BCEMS (at their own station or at a county fire department), fire departments that staffed their own ambulance, fire departments that shared staffing with BCEMS, and even Sibbett's Cab Service, a taxi service in Swannanoa. (Courtesy of Riceville Fire Department.)

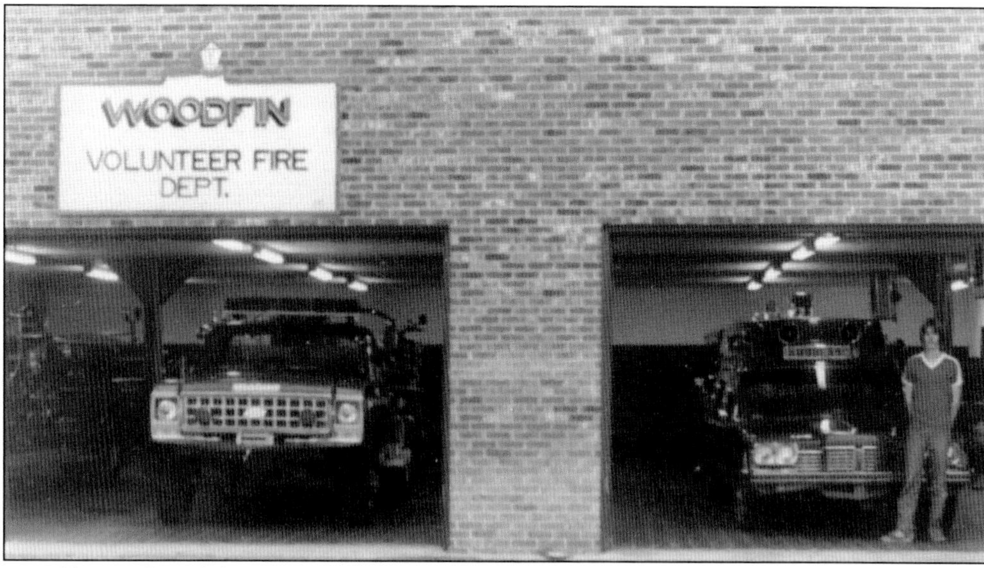

Woodfin purchased this Pontiac ambulance on January 16, 1973, for $15,000, along with $4,000 worth of equipment. It averaged 200 calls per year, with an average annual operating cost of $700. Fire departments came in and out of providing ambulance services. Departments that had an ambulance of some sort include Asheville Stations 1, 3, 4, and 6, Barnardsville, Beaverdam, Black Mountain, Buncombe County Rescue Squad, ENKA, Fairview, French Broad, Leicester, Reems Creek, Reynolds, Riceville, Skyland, Swannanoa, Upper Hominy, Weaverville, West Buncombe, and Woodfin. (Courtesy of Woodfin Fire Department.)

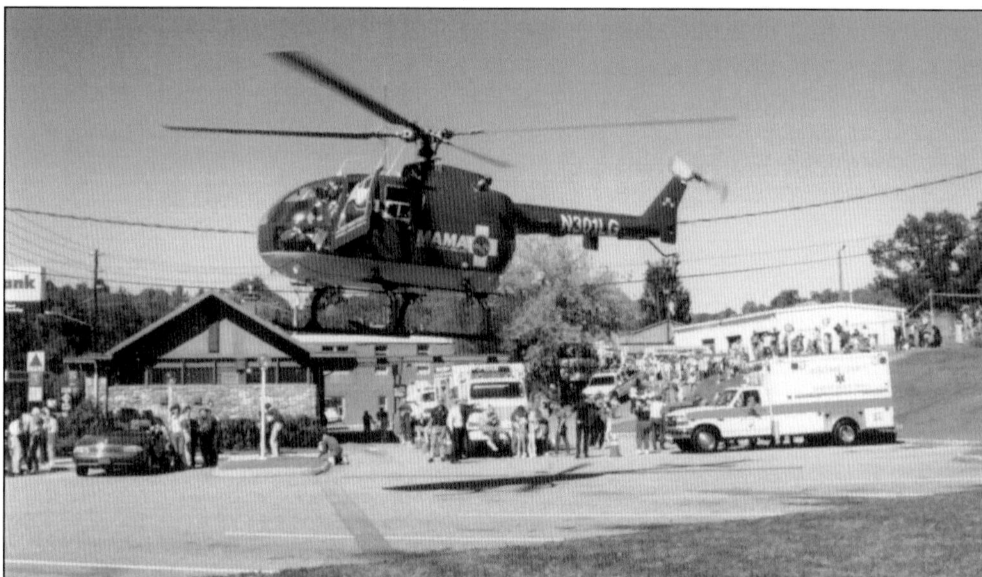

Mission Air Medical Ambulance's (MAMA) first flight occurred on September 29, 1986, in a loaned MBB BO105 helicopter. Starting in March 1987, the above helicopter of the same model helicopter was their first permanent one. In 1997, Mission Air Medical Ambulance became the first civilian program approved for night-vision goggles. In 2004, MAMA 2 was added at Angel Medical Center in Franklin, North Carolina. Since the first flight, more than 19,200 patients have been safely transported. (Courtesy of Skyland Fire Department.)

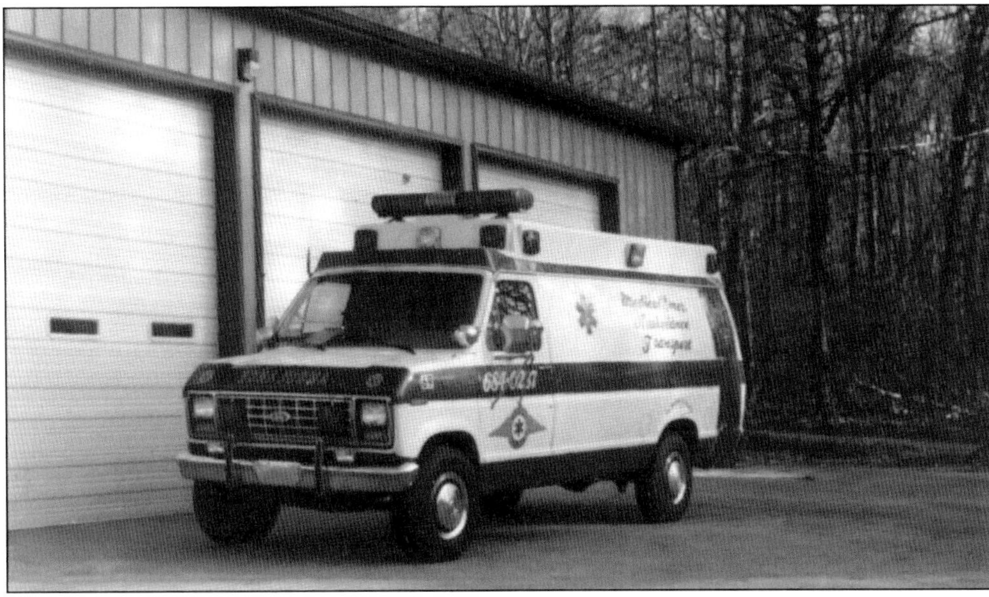

In 1987, Kermit Sr. and Sharon Tolley joined forces with Medical Air Ambulance Transport (MAAT), a privately owned Advanced Lift Support ambulance service. Kermit Sr. retired from BCEMS in 1985, and Sharon and Kermit Jr. used to work for the Penland & Son's Funeral Home ambulance service. The Tolleys took over sole ownership of MAAT in March 1988. It was a tough go, with Kermit and Sharon working for free to try to get business, all patient transfers, from Mission Hospital, nursing homes, private homes, Parkridge Hospital, and the VA medical center. In the spring of 1989, MAAT was closed, and Medical Emergency Ambulance Transport was founded. The name was chosen because it was the cheapest way to relabel the ambulances: easily changing one word. Pictured above and below are the first two Ford ambulances used. (Courtesy of Medic.)

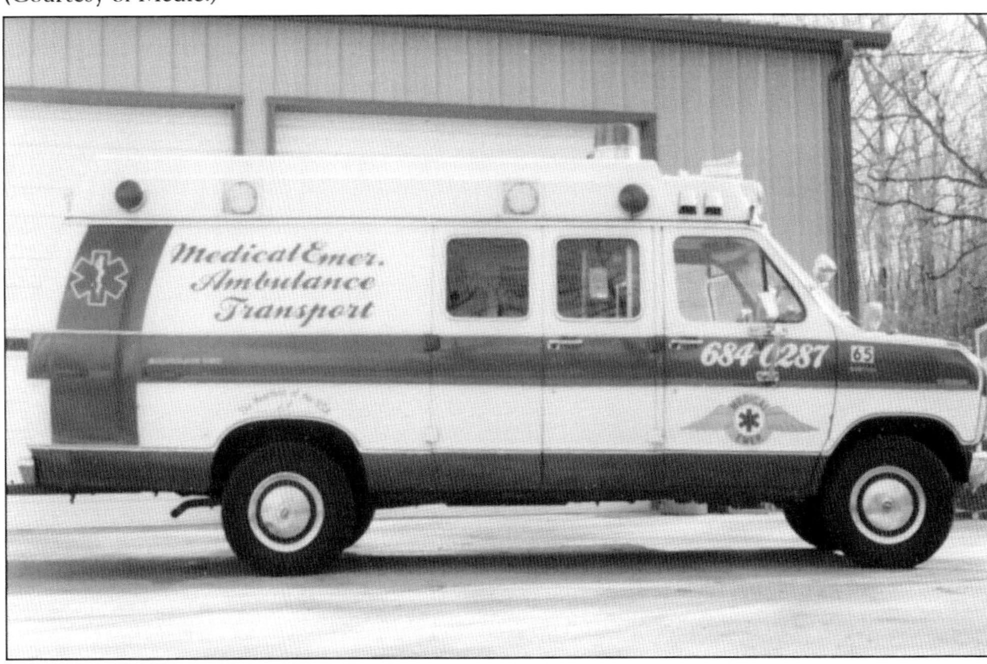

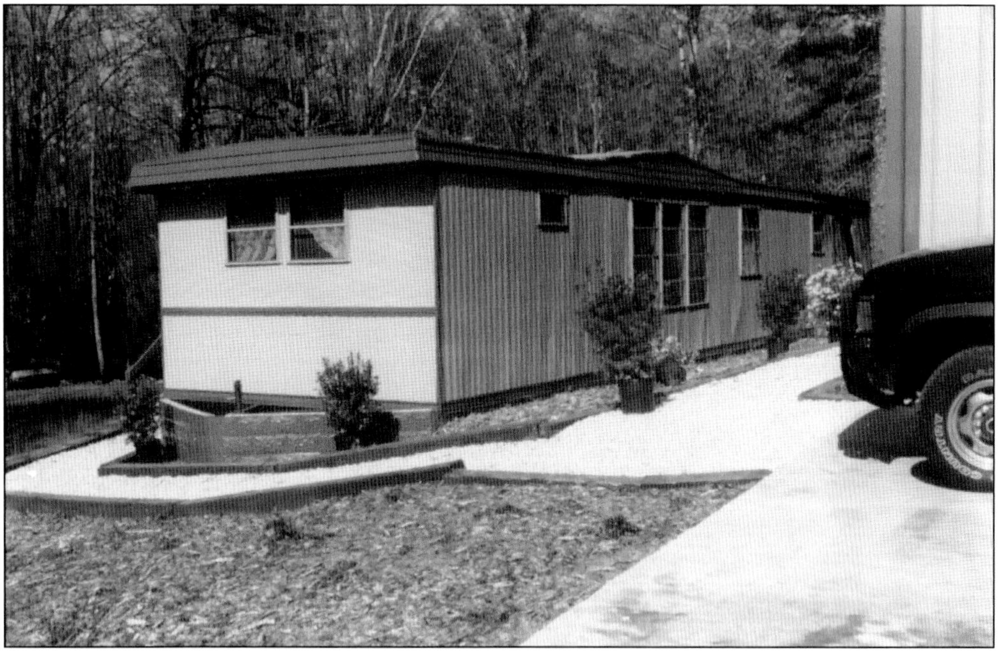

The first office was in the family's beauty shop business in the basement of the family home. The above mobile home was a rental on the family's property before becoming the second office in winter 1989. The ambulances sat outside. The unit numbers first came from MAAT and had to be double numbers because of BCEMS's single numbers. (Courtesy of Medic.)

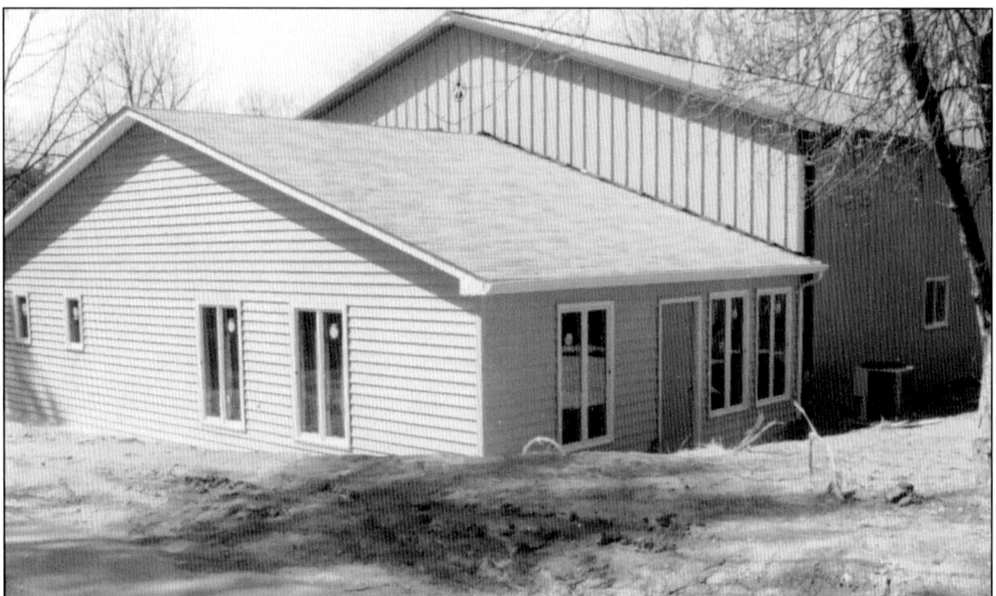

From July to September 1991, the first station for the ambulances was built on the family property shown above. An office addition was built March 1992, and the mobile home was abandoned. In 1993, the name was changed to Medical Emergency Transport, Inc., known as "Medic." Times were tough again in 1993 when MedLine, another private ambulance service, came into town for a year. (Courtesy of Medic.)

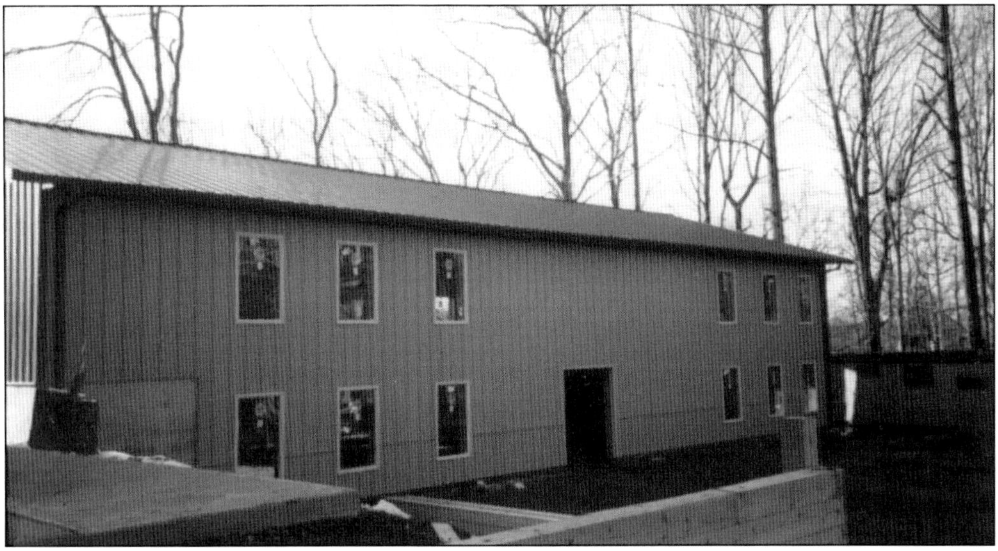

In January 1999, construction began on the third office (shown above), maintenance shop, and race shop. Starting in 2000, medical coverage was provided at Bele Chere, Asheville's free street festival started in 1979 as a way to attract business to the once blighted, now vibrant, downtown. Businesses started abandoning the open-air shopping of old for the new, enclosed Asheville Mall in 1973. It was the Southeast's largest free street festival, attracting around 350,000 attendees each July. The festival name means "beautiful living," and the event celebrated Asheville's incredible local art, food, microbrews, and music scene. The festival ended in 2013. Medic built the below Polaris ambulance for Bele Chere as a way for injured festivalgoers to have some privacy when being transported through the crowded streets. (Both, courtesy of Medic.)

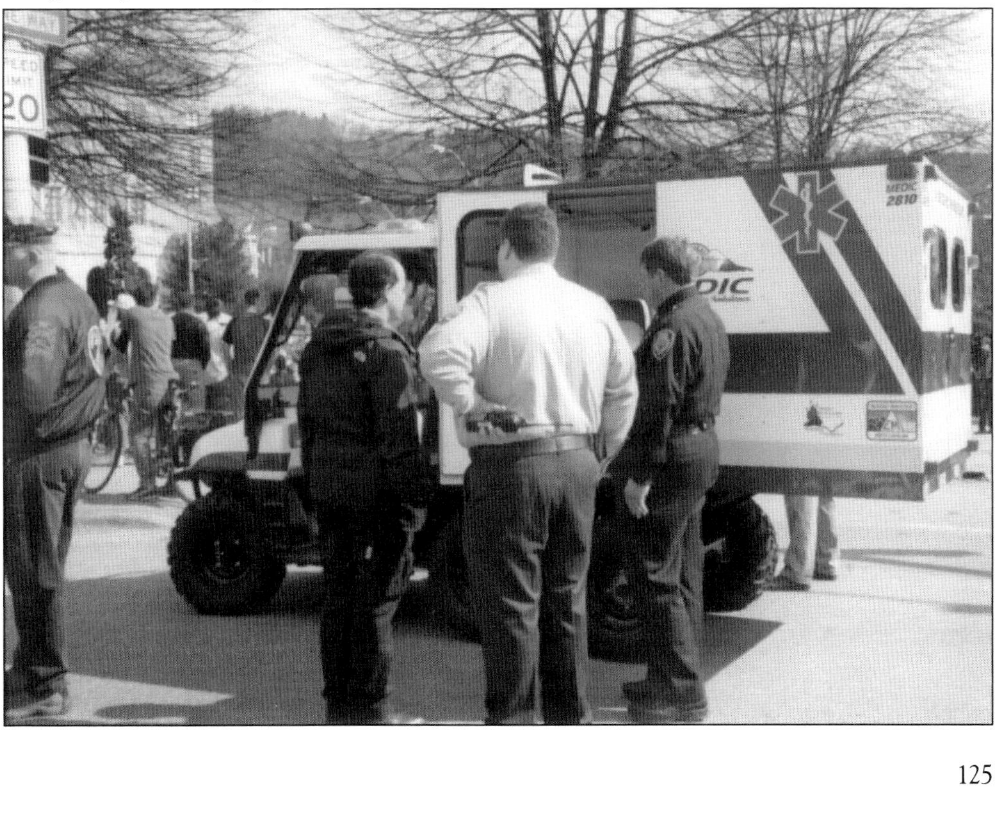

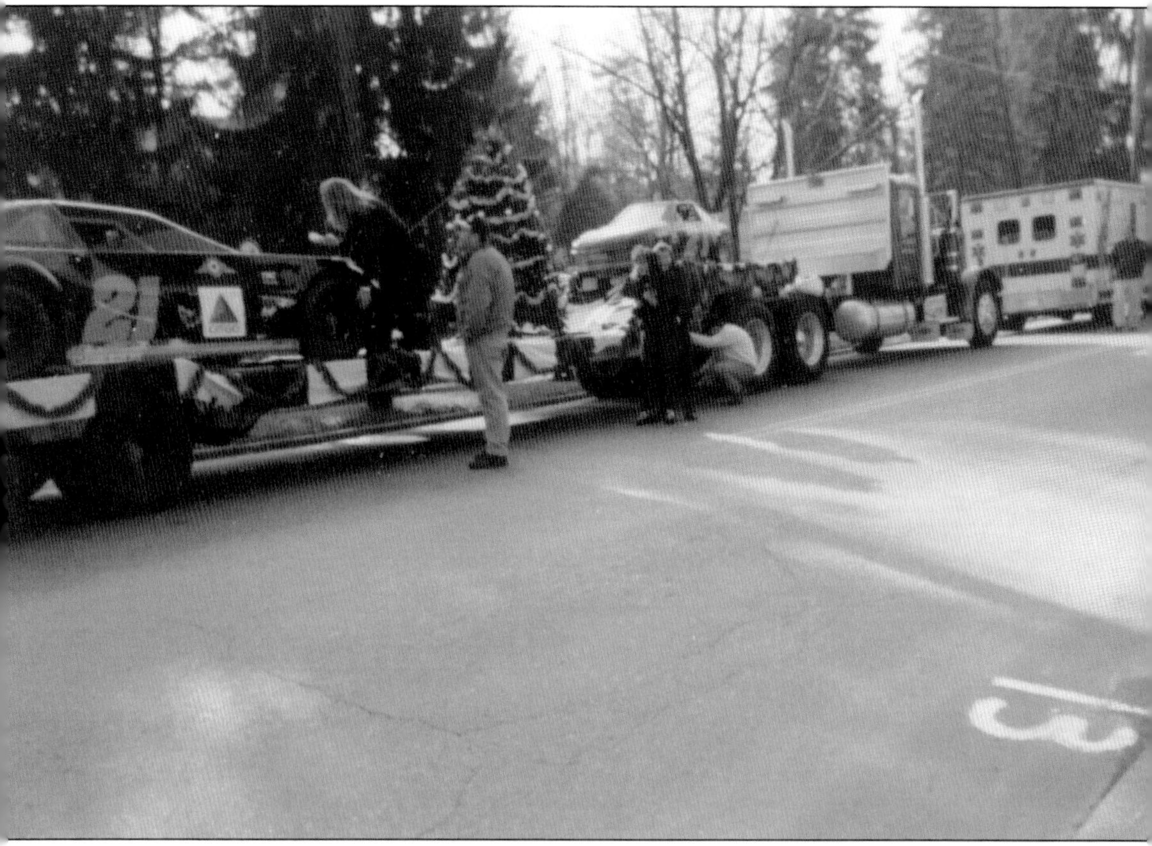

The Tolley family had a love for racing. Kermit Jr. raced late-model stock cars, and his sister Tonya raced the smaller Winston mini, both shown above. Racing was held at the Asheville Motor Speedway on Amboy Road. Medic provided EMS coverage for the races, which were known as Racin' on the River. In 1960, NASCAR came to town with racing legends such as Richard Petty, Junior Johnson, Dale Earnhardt, and the first-ever female winner in a NASCAR touring event, Shawna Robinson. NASCAR left town in 1971, and local racers took over. RiverLink bought the track and donated it to the City of Asheville to build a greenway in 1998. A group of local racers known as the "Speedway 99," which included the Tolleys, was allowed to operate the track for one more year, with the last race being held on September 17, 1999. Medic's day-to-day operations were passed on to Kermit Jr. and Tonya in 2011. They now have eight ambulances on the road. (Courtesy of Medic.)

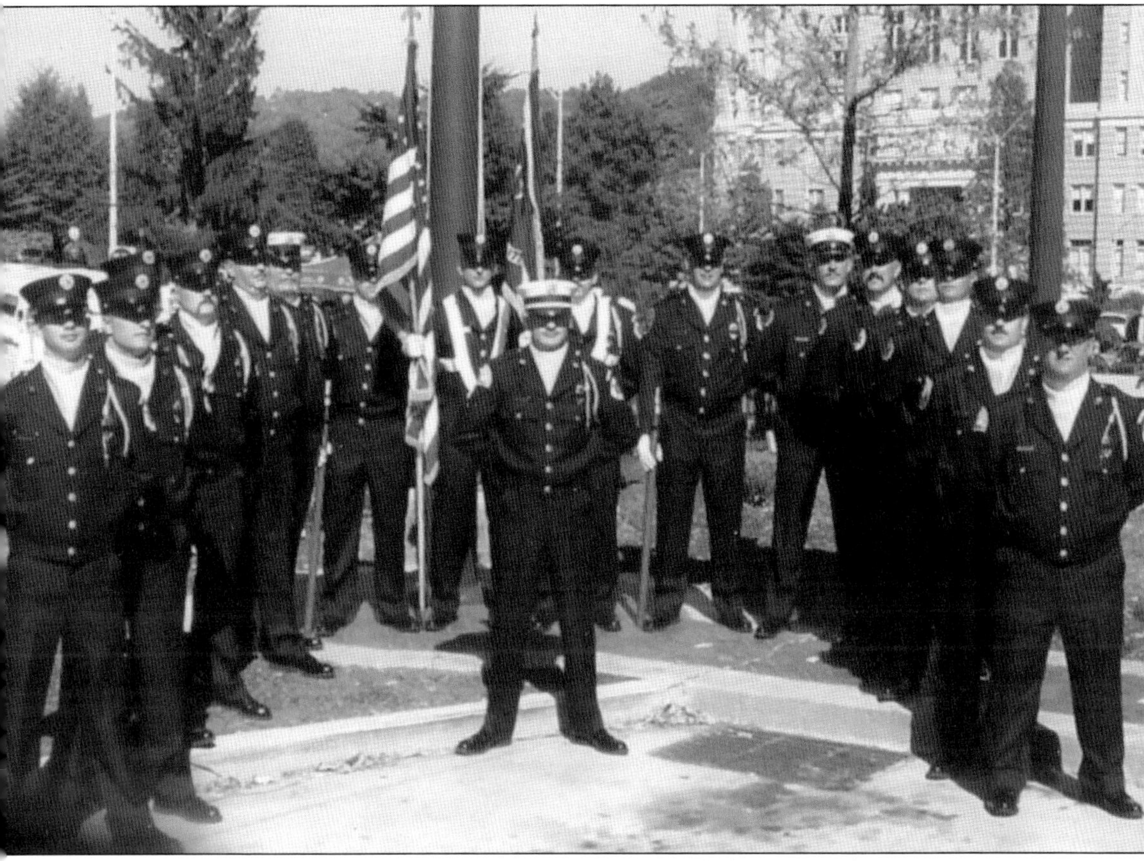

Under the direction of Skyland Fire Department's Rick Wood, the Buncombe County Honor Guard was formed in 1999; it was made up of 24 fire and EMS personnel representing 10 different departments across Buncombe County. Duties included providing pallbearers, casket guards, a full color guard; folding and presenting of the American flag; ringing a fire-service bell; and ensuring honor and respect to fallen firefighters and their families. (Courtesy of the Skyland Fire Department.)